U0260322

中国梨品种

Pear Varieties in China

曹玉芬 主编

Chief Editor Cao Yufen

中国农业出版社

China Agriculture Press

《中国梨品种》编委会

The Editorial Committee

前言

中国是梨属植物的起源中心，也是世界栽培梨的多样性中心之一。梨树在中国已有3 000多年的栽培历史，经过长期的自然选择和人工培育，形成了丰富多样的种质资源，具有东方梨的特色。

梨树在中国分布广泛，除海南省和港、澳地区外均有梨树栽培，栽培面积较大的省、自治区、直辖市包括河北、辽宁、四川、新疆、云南、河南、陕西、贵州、山东、江苏、湖北、安徽、甘肃、山西、重庆、湖南、江西和浙江等。2012年中国梨栽培总面积108.86万hm^2，占世界梨栽培总面积的69.6%；总产量1 707.30万t，占世界总产量的66.5%。主要栽培种有白梨、砂梨、秋子梨和西洋梨等。地方品种繁多，主要为中、晚熟品种；新品种选育成就喜人，相继育成了早酥、黄花、翠冠、黄冠、中梨1号、玉露香等为代表的优良品种100余个，梨鲜果已实现周年供给。

为总结中国梨品种资源，促进我国梨产业的发展，让世界同行对中国梨栽培的种类和品种有所了解，我们编写了《中国梨品种》。全书包括中国梨概况和中国梨栽培品种两部分。收集梨栽培品种174个，每个品种记述来源与分布、主要形态特征和生物学特性以及品质特性等，并附果实田间照片、室内剖面照片及花序照片。该书是比较全面反映中国梨树栽培品种与选育成就的一本科学专著，可以作为梨树科研、育种及生产的参考书。

本书在编写过程中得到了现代农业产业技术体系的支持，许多梨科研单位参与了图片拍摄和样品提供，在此一并致谢。

由于本书编写时间较短，收集到的资料不够全面，有待以后进一步补充完善。此外，限于编著者水平，难免存在疏漏，敬请读者批评指正。

编　者

2014年6月

FOREWORD

China is the origin center of plants in genus *Pyrus* and one of the diversity centers for cultivated pear trees in the world. With an over 3000-year history of cultivation and long-term natural selection and artificial improvement, China has abundant pear germplasm resources with obvious oriental pear characteristics.

Pear tree is widely distributed in entire China except Hainan Province, Hong Kong and Macao, and mostly cultivated in the provinces (autonomous regions or municipalities) of Hebei, Liaoning, Sichuan, Xinjiang, Yunnan, Henan, Shaanxi, Guizhou, Shandong, Jiangsu, Hubei, Anhui, Gansu, Shanxi, Chongqing, Hunan, Jiangxi, Zhejiang, etc. In 2012, the pear tree acreage and annual pear production in China reached 1088.6 thousand hectares and 17.073 million tons, respectively, accounting for 69.6% of the total pear tree planting area and 66.5% of the annual pear production, respectively, in the world. The principle pear cultivars are White Pear, Sand Pear, Ussurian Pear, and Common Pear, etc, with abundant local accessions, mainly mid or late matured varieties. Pear improvement obtained gratifying achievement with the release of more than 100 new pear cultivars, such as Zaosu, Huanghua, Cuiguan, Huangguan, Zhongli No.1, Yuluxiang, etc. and pear fruits able to meet the all year-round market demand.

To promote pear industry in China and introduce more Chinese pear varieties to the worldwide pear community, we organized a committee to edit and publish this book *Pear Varieties in China*. The book consists of two parts: the general pear situation and the cultivated pear varieties in China, and introduces 174 cultivars with their origin, distribution as well as main morphological, biological and quality characteristics. In addition, the book also provides typical pictures of pear fruits in the orchards, cross and longitudinal sections of indoor pear fruits, as well as the inflorescence. Overall, this book is the scientific monograph that wholly introduces pear varieties in China and could be used as a reference for pear research, breeding and planting.

At last, we would like to take this opportunity to thank the earmarked fund for China Agriculture Research System for its financial support and research institutes engaged in pear research for providing samples and photographing. Due to the limited time and abilities, the book needs further supplement in the future. Criticism and rectification to the careless omissions and faultiness are respected and appreciated.

Editors
June, 2014

目录 CONTENTS

第一部分
中国梨概况

Part Ⅰ General Situation of Pear Industry in China

一、中国梨简况

Brief Introduction

（一）中国梨栽培历史悠久 A Long History of Pear Cultivation in China

中国是世界栽培梨历史最悠久的国家之一。其栽培究竟始于何时已无法考证，但在先秦时代，《诗经》就有“山有苞棣，隰有树檖”的记载。《诗经》的秦风章“晨风”篇作于公元前1000年左右，可见有文献记载的栽培历史就有3 000多年。此后的《山海经》《庄子》《韩非子》《尔雅》《魏书》等古书中均有关于梨的记载。长沙马王堆汉墓中发现了梨核及记载了梨树种植情况的竹简，说明西汉时期梨就有大量栽培，且有了很多品种，甚至有了防治虫害的方法。魏晋时期，梨树的栽培与繁育有了长足的发展，北魏贾思勰所著的《齐民要术》不仅记载了多个梨树品种，还详细地记载了梨树的栽培技术，总结了梨树的嫁接技术和贮藏技术。丰富的古代文献记载及诗词佐证了中国悠久的梨栽培历史。

China is one of the countries with the longest history of pear cultivation in the world. Although the exact time when pears were firstly cultivated in China can not be verified, record about a kind of pear tree named ‘Sui’ was found in the article “ChenFeng” in chapter “QinFeng”, book of *ShiJing* published in the pre-Qin era around 1000 BC. Thus, the history of pear cultivation in China is speculated to be about three thousand years. The book *Shan Hai Jing*, *Zhuang Zi*, *Han Feizi*, *ErYa*, *Wei History* and other late ancient books all have records about pears. Pear cores and bamboo with documentation of pear cultivation were found in the tomb of Han Dynasty in Changsha Mawangdui, indicating that pears had been widely cultivated in Western Han Dynasty, and there were many pear varieties and even pest controlling ways at that time. Pear cultivation and breeding had made great progress during Wei and Jin Dynasties. The book *Qi Min Yao Shu* written by Jia Sixie in Wei Dynasty documented not only a number of pear varieties, but also detailed techniques about pear cultivation, and summarized pear grafting techniques and storage technologies. Abundant evidences of ancient literature and poetry corroborated that China has a long history of pear cultivation.

（二）中国是梨的起源中心及栽培梨的多样性中心之一 China Is the Origin Center and One of Diversity Centers for Cultivated Pears

中国是梨的原生中心。梨起源于第三纪中国西部的山区，分布有丰富的梨亚科类型，有世界梨资源的宝库之美誉。梨树从起源地沿着山脉分别向东和向西传播，形成了中国、中亚和近东三个次生中心。中国次生中心分布有砂梨、秋子梨、豆梨、杜梨等东方梨的代表种，其中砂梨野生于中国的长江流域，秋子梨野生于中国的东北、华北北部、西北各地，豆梨野生于华东、华南各地，杜梨野生于华北、西北各地，以及湖北、江苏、安徽等地，演化成适合于不同生态条件的变种或生态类型，广泛分布具有丰富遗传多样性的栽培地方品种。

China is the origin center of pear. The genus *Pyrus* is believed to have arisen during the Tertiary period in the mountainous regions of western China, which are now rich in forms of Pomoideae, and honored as a treasure trove of pear resources in the world. Pear germplasm had spread from the center along the mountain chains both eastward and westward and formed three secondary centers in China, Central Asia and the Near East. Pear germplasm in the secondary center in China includes the representative species of oriental pears such as Sand Pear [*P. pyrifolia* (Burm.f.) Nakai], Ussurian Pear (*P. ussuriensis* Maxim.), Callery Pear (*P. calleryana* Dcne.) and Birch-leaf Pear (*P. betulaefolia* Bge.), etc. Among them, *P. pyrifolia* is widely distributed in Yangtze River basin; *P. ussuriensis* in Northeast, North and Northwest China; *P. calleryana* in East and South China; *P. betulaefolia* in North, and Northwest China, as well as in Hubei, Jiangsu, Anhui provinces. These species evolved to abundant local pear varieties with broad genetic diversities and ecotypes suitable for different ecological conditions.

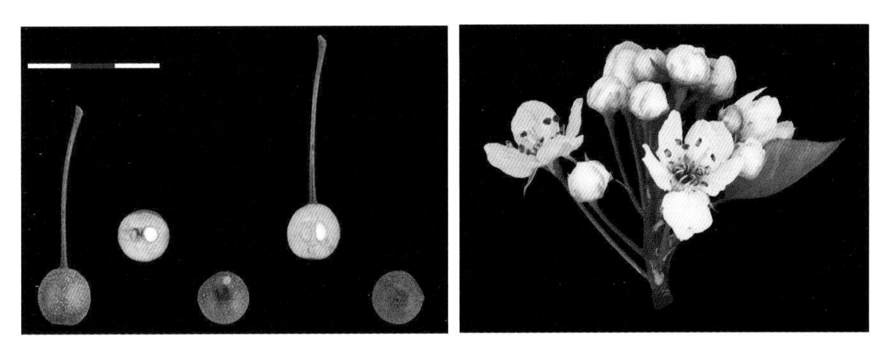

豆梨 Callery Pear

杜梨 Birch-leaf Pear

秋子梨 Ussurian Pear

黑龙江野生山梨 Wild Ussurian Pear in Heilongjiang

云南野生川梨 Wild Pashi Pear in Yunnan

山西野生杜梨 Wild Birch-leaf Pear in Shanxi

（三）中国梨种质资源丰富 China Is Rich in Pear Germplasm

作为梨的原生地和东方梨的分化中心，中国拥有非常丰富的梨属植物资源。《中国果树志·梨》描述了白梨类品种459个、砂梨类452个、秋子梨类72个、新疆梨类29个、川梨类10个，此外还有部分西洋梨及日本梨品种。目前，中国已经安全保存梨种质2 000余份（含部分重复），主要保存在位于辽宁、湖北、吉林、云南、新疆的5个国家果树种质资源圃中，包含了野生类型、半栽培类型、地方品种、选育品种、遗传材料及国外引进品种。

As the origin and diversity center of oriental pears, China is very rich in *Pyrus* germplasm. The book *Chinese Fruit-Pear* described 459 varieties of Chinese White Pear (*P. bretschneideri* Rehd.), 452 varieties of Sand Pear, 72 varieties of Ussurian Pear, 29 varieties of Sinkiang Pear (*P. sinkiangensis* Yu) and 10 varieties of Pashi Pear (*P. pashia* Buch.-Ham. ex D. Don) as well as some of the Japanese and Common Pear varieties. Nowadays, China has safely preserved more than 2 000 accessions of pear germplasm including wild accessions, semi-cultivated accessions, local varieties, breeding cultivars, genetic materials and varieties introduced from abroad mainly in the 5 national fruit germplasm repositories located in Liaoning, Hubei, Jilin, Yunnan and Xinjiang.

（四）中国是梨生产大国 China Is the Largest Pear Producing Country

中国是世界梨第一生产大国。2012年中国梨栽培总面积和总产量分别为108.86万hm^2和1707.30万t，占全国水果总面积和总产量的8.96%和11.30%，占世界梨总面积和总产量的69.6%和66.5%。最近10年中国梨栽培面积基本稳定，但总产量增加了近40%，主要是由于新建梨园进入盛果期，以及新品种、新技术和新模式的推广应用，使得单产显著提高。各省份梨生产情况如表1。中国梨出口规模也逐年扩大，并于2009年超越阿根廷成为全球第一梨出口大国，2010年出口43.79万t，占世界总出口量的17.05%。

China is the largest pear producer in the world. The total cultivated area and production of pear in China was 1 088 600 hectares and 17.073 million tons, respectively, accounting for 8.96% and 11.30% of the total area and fruit production of China as well as 69.6% and 66.5% of the total pear planting area and production in the world (2012 data). The cultivated area of pear in China is basically static during the last 10 years, but the production increased nearly 40% mainly because of significantly improved pear yield per unit area resulted from the use of new varieties, new technologies and new planting molds, as well as new orchards entering high yielding stage. The pear production in each province is shown in Table 1. In addition, China's pear export

表1 中国梨栽培面积与产量（2012）

Table 1 Pear cultivated area and production in China (2012)

省（自治区、直辖市） Province(autonomous region and municipality)		面积（khm^2） Area (thousand hectare)	产量（t） Production (ton)
北　京	Beijing	9.07	162 632
天　津	Tianjin	4.11	36 218
河　北	Hebei	193.97	4 450 544
山　西	Shanxi	35.10	663 588

（续）

省（自治区、直辖市） Province(autonomous region and municipality)		面积（km^2） Area (thousand hectare)	产量（t） Production (ton)
内蒙古	Inner Mongolia	7.49	74 924
辽　宁	Liaoning	98.77	1 547 193
吉　林	Jilin	13.71	112 603
黑龙江	Heilongjiang	3.98	37 259
上　海	Shanghai	1.92	37 359
江　苏	Jiangsu	39.42	748 219
浙　江	Zhejiang	23.72	390 500
安　徽	Anhui	37.31	1 069 300
福　建	Fujian	22.00	205 745
江　西	Jiangxi	27.14	140 594
山　东	Shandong	42.48	1 190 939
河　南	Henan	51.99	1 043 927
湖　北	Hubei	37.35	536 352
湖　南	Hunan	32.32	154 253
广　东	Guangdong	7.78	77 982
广　西	Guangxi	21.25	257 690
重　庆	Chongqing	34.95	340 983
四　川	Sichuan	83.30	960 290
贵　州	Guizhou	48.05	217 178
云　南	Yunnan	52.25	416 326
西　藏	Xizang	0.09	1 150
陕　西	Shaanxi	48.58	896 932
甘　肃	Gansu	36.33	333 281
青　海	Qinghai	0.88	4 708
宁　夏	Ningxia	2.04	14 161
新　疆	Xinjiang	70.23	950 197
总　计	Total	1 088.58	17 073 026

scale has increased year by year in past several years. China has replaced Argentina as the world's largest pear exporter in 2009 and exported 437 900 tons of pears in 2010, accounting for 17.05% of the total export volume in the world.

二、中国梨的分布与区划布局

Planting Distribution and Layout of Pears in China

（一）分布区域广泛　Extensive Distribution

我国由南到北，从东到西，除海南省和港、澳地区外均有梨树栽培，是分布最广的果树树种。秋子梨主要分布在辽宁、吉林，白梨主要分布在黄河以北到长城一带，长江以南为砂梨的分布区，黄河、秦岭以南，长江、淮河以北是砂梨和白梨的混交带，西洋梨主要在山东的胶东半岛栽培。2012 年我国栽培面积最大的十个省份梨果产量占到全国总产量的 79% 以上，其中河北省是第一生产大省，占全国总面积的 17.8%，全国总产量的 26.1%。

Pear is one of the most widely distributed fruit trees in China. It is cultivated in the whole country except Hainan Province, Hong Kong and Macao. Among the different kinds of pears, Ussurian Pear is mainly distributed in Liaoning and Jilin Provinces, White Pear is mainly distributed from the areas north to the Yellow River to the areas of the Great Wall, Sand Pear is mainly distributed in the areas south to the Yangtze River, and mixed with White Pear in the areas south to the Qinling Mountains/the Yellow River and north to the Yangtze River/the Huaihe River. Common Pear is mainly cultivated in the Shandong Peninsula. According to the statistical data of 2012, the production of the ten provinces with the largest cultivated area of pear accounted for more than 79% of national output. Hebei Province is the largest pear production province, producing 17.8% and 26.1% of pear in acreage and production, respectively, in China.

（二）传统地方品种繁多，栽培面积较大 Local Varieties Are Abundant and Occupy Large Proportion of the Cultivated Areas in China

中国梨的栽培品种涵盖了白梨、砂梨、秋子梨、新疆梨和西洋梨 5 个种，大面积栽培的品种就达 100 多个。传统的主栽品种有砀山酥梨、鸭梨、南果梨、京白梨、库尔勒香梨、雪花梨、苍溪雪梨、三花梨等，均为古老地方品种，成熟期中或晚。在品种构成上，一直以传统地方品种为主，砀山酥梨、鸭梨、雪花梨等晚熟品种约占全国总产量的 40%。

Pear varieties in China belong to five species: White Pear, Sand Pear, Ussurian Pear, Sinkiang Pear and Common Pear. Among them, more than 100 varieties are cultivated with large areas. The major traditional cultivars are ancient local varieties with late or mid maturity including Dangshan Suli, Yali, Nanguoli, Jingbaili, Korla Pear, Xuehuali, Cangxi Xueli, Sanhuali and so on. Local varieties account for majority of the cultivated pears. Dangshan Suli, Yali, Xuehuali and other late-maturing varieties constitute about 40% of the total cultivation composition in China.

表2 中国梨主要栽培品种与砧木

Table 2 Main cultivars and rootstocks of pear in China

省（自治区、直辖市） Province (autonomous region and municipality)	主栽品种 Main cultivars	主要砧木 Main rootstocks
北 京 Beijing	鸭梨、黄金梨、丰水、雪花梨 Yali, Whangkeumbae, Housui, Xuehuali	杜梨 *P. betulaefolia*
天 津 Tianjin	鸭梨、雪花梨、黄冠 Yali, Xuehuali, Huangguan	杜梨 *P. betulaefolia*
河 北 Hebei	鸭梨、雪花梨、黄冠、中梨1号 Yali, Xuehuali, Huangguan, Zhongli No.1	杜梨 *P. betulaefolia*
山 西 Shanxi	砀山酥梨、玉露香、红香酥、中梨1号、高平黄梨、晋蜜梨 Dangshan Suli, Yuluxiang, Hongxiangsu, Zhongli No.1, Gaoping Huangli, Jinmili	杜梨 *P. betulaefolia*
内蒙古 Inner Mongolia	南果梨、苹果梨 Nanguoli, Pingguoli	山梨 *P. ussuriensis*

（续）

省（自治区、直辖市） Province (autonomous region and municipality)	主栽品种 Main cultivars	主要砧木 Main rootstocks
辽　宁 Liaoning	南果梨、苹果梨、秋白梨、锦丰、尖把梨、鸭梨、巴梨、花盖、早酥 Nanguoli, Pingguoli, Qiubaili, Jinfeng, Jianbali, Yali, Bartlett, Huagai , Zaosu	山梨、杜梨 *P. ussuriensis* *P. betulaefolia*
吉　林 Jilin	苹果梨、南果梨、苹香梨 Pingguoli, Nanguoli, Pingxiangli	山梨 *P. ussuriensis*
黑龙江 Heilongjiang	苹果梨、金香水、龙园洋红，龙园洋梨，秋香 Pingguoli, Jinxiangshui, Longyuan Yanghong, Longyuan Yangli, Qiuxiang	山梨 *P. ussuriensis*
上　海 Shanghai	翠冠、清香 Cuiguan, Qingxiang	杜梨、豆梨 *P. betulaefolia* *P. calleryana*
江　苏 Jiangsu	砀山酥梨、翠冠 Dangshan Suli ， Cuiguan	杜梨 *P. betulaefolia*
浙　江 Zhejiang	翠冠、黄花、清香、翠玉 Cuiguan, Huanghua, Qingxiang, Cuiyu	杜梨、豆梨 *P. betulaefolia* *P. calleryana*
安　徽 Anhui	砀山酥梨、黄冠、翠冠、徽州雪梨 Dangshan Suli， Huangguan, Cuiguan, Huizhou Xueli	杜梨 *P.betulaefolia*
福　建 Fujian	翠冠、黄花、台农2号、新世纪 Cuiguan, Huanghua, Tainong No.2, Shinseiki	豆梨、杜梨 *P. calleryana* *P. betulaefolia*
江　西 Jiangxi	翠冠、黄花、清香 Cuiguan, Huanghua, Qingxiang	豆梨、杜梨 *P. calleryana* *P. betulaefolia*
山　东 Shandong	鸭梨、新高、茌梨、黄金梨、长把、巴梨 Yali, Niitaka, Chili, Whangkeumbae, Changba, Bartlett	杜梨 *P. betulaefolia*
河　南 Henan	砀山酥梨、黄金梨、圆黄、中梨1号、翠冠、秋黄梨 Dangshan Suli, Whangkeumbae, Wonhwang, Zhongli No.1, Cuiguan, Chuwhangbae	杜梨、豆梨 *P. betulaefolia* *P. calleryana*
湖　北 Hubei	湘南、黄花、华梨1号、翠冠、圆黄 Shounan, Huanghua, Huali No.1, Cuiguan, Wonhwang	豆梨、杜梨 *P. calleryana* *P. betulaefolia*
湖　南 Hunan	金秋梨、黄花、翠冠 Jinqiuli, Huanghua, Cuiguan	豆梨、杜梨 *P. calleryana* *P. betulaefolia*
广　东 Guangdong	翠冠、黄花、新世纪、洞冠梨 Cuiguan, Huanghua, Shinseiki, Dongguanli	豆梨 *P. calleryana*
广　西 Guangxi	灌阳雪梨 Guanyang Xueli	豆梨 *P. calleryana*
重　庆 Chongqing	黄花、翠冠、中梨1号 Huanghua, Cuiguan, Zhongli No.1	杜梨、川梨 *P. betulaefolia* *P. pashia*
四　川 Sichuan	翠冠、苍溪梨、黄花、金花梨、黄金梨 Cuiguan, Cangxili, Huanghua, Jinhuali, Whangkeumbae	杜梨、川梨 *P. betulaefolia* *P. pashia*
贵　州 Guizhou	金秋梨，威宁大黄梨 Jinqiuli, Weining Dahuangli	豆梨 *P. calleryana*

（续）

省（自治区、直辖市） Province (autonomous region and municipality)	主栽品种 Main cultivars	主要砧木 Main rootstocks
云　南 Yunnan	美人酥、满天红、早白蜜、金花梨、早酥 Meirensu, Mantianhong, Zaobaimi, Jinhuali, Zaosu	杜梨、川梨 *P. betulaefolia* *P. pashia*
西　藏 Xizang		
陕　西 Shaanxi	砀山酥梨、早酥、红香酥、黄金梨 Dangshan Suli, Zaosu, Hongxiangsu, Whangkeumbae	杜梨 *P. betulaefolia*
甘　肃 Gansu	早酥、苹果梨、黄冠、冬果梨、软儿梨、皮胎果、巴梨 Zaosu, Pingguoli, Huangguan, Dongguoli, Ruan'erli, Pitaiguo, Bartlett	杜梨、木梨 *P. betulaefolia* *P. xerophila*
青　海 Qinghai	早酥 Zaosu	杜梨 *P. betulaefolia*
宁　夏 Ningxia	早酥 Zaosu	杜梨 *P. betulaefolia*
新　疆 Xinjiang	库尔勒香梨、砀山酥梨、苹果梨 Korla Pear, Dangshan Suli, Pingguoli	杜梨 *P. betulaefolia*

（三）新品种选育成就喜人，品种趋于多样化，成熟期延长 Gratifying Achievements in Breeding New Varieties with Diversified Characters and Effectively Extended Maturity

砀山酥梨、鸭梨等传统的主栽品种为中国梨产业做出了巨大贡献，而且还将继续发挥其重要作用。自20世纪50年代起，中国有计划、系统、科学地开展了梨品种选育工作。尽管育种工作起步较晚，但经过几代人的努力，凭借拥有丰富梨种质资源优势，新品种选育工作取得了骄人业绩。相继育成了以早酥、黄花、翠冠、黄冠、中梨1号、玉露香等为代表的优良品种100余个。中国农业科学院果树研究所育成的早酥，作为中国梨育种的标志性品种，在白梨产区发挥了重要作用，有力地推动了中国梨产业发展。80～90年代，原浙江农业大学育成的黄花成为砂梨的标志性品种。此后，石家庄果树研究所育成了黄冠，浙江省农业科学院园艺研究所育成了翠冠，由于这两个品种综合性状优异，品种育成后就快速在白梨和砂梨产区大规模推广应用，形成了梨新品种“南有翠冠，北有黄冠”的格局。除了上述育种单位外，全国专业开展梨育种的单位有20余家，从事育种工作的科研人员近百人，成立了全国梨育种协作组。我国已成为全世界从事梨育种工作人员最多的国家，也是育成品种最多的国家。由于早熟、特早熟品种育成与推广，梨鲜果采收期延长了近两个月，结合贮藏保鲜，梨鲜果已实现了周年供应。

The traditional cultivars represented by Dangshan Suli and Yali have been making tremendous contribution to the pear industry in China. After the 1950s, China carried out pear breeding plans. Although the work started relatively late, remarkable achievements have been made by the efforts of several generations of breeders on the basis of rich pear resources. More than 100 varieties including Zaosu, Huanghua, Cuiguan, Huangguan, Zhongli No.1 and Yuluxiang, etc., have been released during this period. The pear cultivar (cv.) Zaosu, bred by the Fruit Research Institute, Chinese Academy of Agricultural Sciences and regarded as the symbol of Chinese White Pear varieties, has played an important role in White Pear production and strongly promoted the development of Chinese pear industry. From 1980s to 1999s, the cv. Huanghua bred by the former Zhejiang Agricultural University became a representative variety of Sand Pear. Later on, the cv. Huangguan and cv. Cuiguan were bred by the Shijiazhuang Institute of Pomology and the Institute of Horticulture of Zhejiang Academy of Agricultural Sciences, respectively. Due to the excellent comprehensive quality, the two varieties were spread rapidly and largely planted in the Sand Pear and White Pear planting regions, forming a new planting pattern named "South Cuiguan and North Huangguan" in China. Except for these mentioned above, more than 20 other professional groups of nearly 100 researchers engaged in pear breeding were united to establish the National Pear Breeding Cooperative Group. Currently, China has not only the most staff engaged in pear breeding, but also the most self-bred varieties in the world. The harvest period of pear fruits was extended nearly two months by planting newly bred varieties with extremely early or early maturity. Pear fruits could be supplied all the year round in Chinese market since promotion of new varieties and advancement in storage and preservation technology.

（四）梨产业优势区域布局逐渐形成 Gradual Formation of Regionally Distributed Favorable Areas

经过长期的自然选择和生产发展，逐步形成了秋子梨、白梨及砂梨等的优势产区。2009年，根据市场需求、生态条件、产业基础三大原则条件，农业部颁布了全国梨重点区域发展规划（2009—2015年），将传统的梨优势产区划分为"三区四点"。即华北白梨区、西北白梨区和长江中下游砂梨区三区；辽宁南部鞍山和辽阳的南果梨重点区域、新疆库尔勒和阿克苏的库尔勒香梨重点区域、云南泸西和安宁的红梨重点区域、胶东半岛西洋梨重点区域四个点。在"三区四点"中，河北的鸭梨和雪花梨、安徽的砀山酥梨、新疆的库尔勒香梨、辽宁的南果梨都是全国乃至世界知名的梨品种，具有区域特色和生产优势。此外，还有少量从国外引进的二十世纪、丰水、圆黄、巴梨、康佛伦斯等品种。近年来，翠冠、黄冠等早中熟新品种发展迅速，尤其是长江流域及其以南地区的南方早熟梨蓬勃发展，早熟品种比例显著增加，早、中、晚熟品种结构趋向合理，形成了以优良地方品种和自主育成的梨新品种为主、引进品种为辅的良好品种结构。

Favorable areas of Ussurian Pear, White Pear and Sand Pear have formed during the long-term natural selection and production development. According to the three principles of market requirements, ecological conditions and industrial conditions, the Ministry of Agriculture of China issued a nationwide plan for key pear development regions (2009－2015). The traditional pear favorable areas were classified as "*three zones and four key areas*", i.e. the White Pear Zone of Northern China, the White Pear Zone of Northwestern China and the Sand Pear Zone along the middle and lower reaches of Yangtze River; the Key Area of Nanguoli in Anshan and Liaoyang, south of Liaoning Province, the Key Area of Korla Pear in Korla and Aksu of Xinjiang, the Key Area of Red Pear in Luxi and Anning of Yunnan Province and the Key Area of Common Pear in Shandong Peninsula. In the "*three zones and four key areas*", Yali and Xuehuali in Hebei Province, Dangshan Suli in Anhui Province, Korla Pear in Xinjiang Uygur Autonomous Region and Nanguoli in Liaoning Province are national and even world-renowned pear varieties with regional characteristics and production advantages. There are also a small amount of introduced varieties including Nijisseiki, Housui, Wonhwang, Bartlett, and Conference, etc. In recent years, early maturing varieties including Cuiguan and Huangguan have developed rapidly, especially in the areas along and south to the Yangtze River. The proportion of early maturing varieties has increased significantly and the proportion of early, middle and late-maturing varieties is becoming more rational, forming a fine variety composition of dominant local varieties and self-bred new varieties and supplemental introduced varieties.

甘肃皋兰县古梨园
Old pear orchard in Gaolan County of Gansu Province

杭州平原地区棚架翠冠、黄花梨园
Trellis orchard of Cuiguan and Huanghua in plain of Hangzhou

新疆库尔勒市库尔勒香梨园
Korla Pear orchard in Korla City of Xinjiang

福建建宁山地黄花梨园
Huanghua pear orchard on mountainous area in Jianning County of Fujian Province

浙江松阳山地翠冠梨园
Cuiguan pear orchard on mountainous area in Songyang County of Zhejiang Province

河北省泊头市鸭梨园
Yali orchard in Botou City of Hebei Province

辽宁海城南果梨园
Nanguoli orchard in Haicheng City of Liaoning Province

吉林龙井市苹果梨园
Pingguoli orchard in Longjing City of Jilin Province

安徽砀山县砀山酥梨园
Dangshan Suli orchard in Dangshan County of Anhui Province

河北高阳县雪青、黄冠梨园
Xueqing and Huangguan pear orchard in Gaoyang County of Hebei Province

甘肃景泰早酥梨密植园
Zaosu pear density orchard in Jingtai County of Gansu Province

山西太谷玉露香梨园
Yuluxiang pear orchard in Taigu County of Shanxi Province

第二部分
中国梨栽培品种

Part II Cultivated Varieties of Pear in China

一、中国梨地方品种

Local Pear Varieties in China

（一）白梨品种 White Pear Varieties

1. 博山池梨

来源及分布 2n=34，原产山东省。

主要性状 树势强，丰产，始果年龄晚。叶片长 11.7cm，宽 8.7cm，卵圆形或阔卵圆形，初展叶绿色，着红色。花蕾白色，每花序 5 ～ 7 朵花，平均 5.7 朵；雄蕊 19 ～ 21 枚，平均 20.0 枚；花冠直径 3.7cm。在辽宁兴城，果实 9 月下旬至 10 月上旬成熟，单果重 156g，纵径 6.8cm，横径 6.7cm，倒卵圆形；果皮绿色或黄绿色；果心中大，5 心室；果肉白色，肉质细，松脆，汁液多，味淡甜；含可溶性固形物 12.10%，可滴定酸 0.15%；品质中上等。果实耐贮藏。

1. Boshan Chili

Origin and Distribution Boshan Chili (2n=34), was originated in Shandong Province.

Main Characters Tree: vigorous, productive, late precocity. Leaf: 11.7cm × 8.7cm in size, ovate or broadly ovate. Initial leaf: green tinged with red. Flower: white bud, 5 to 7 flowers per cluster in average of 5.7; stamen number: 19 to 21, averaging 20.0; corolla diameter: 3.7cm. Fruit: matures in late September or early October in Xingcheng, Liaoning Province, 156g per fruit, 6.8cm long, 6.7cm wide, obovate, green or yellowish-green skin, medium core, 5 locules, flesh white, fine, crisp tender, juicy, light sweet; TSS 12.10%, TA 0.15%; quality above medium; storage life long.

2. 槎子梨

来源及分布 原产山东省平邑天宝山区，在山东平邑、费县、滕县等地有栽培。

主要性状 树势中庸，丰产，始果年龄中等。叶片长 10.4cm，宽 7.7cm，卵圆形，初展叶绿色，着红色。花蕾白色，边缘浅粉红色，每花序 5 ~ 7 朵花，平均 6.0 朵；雄蕊 18 ~ 22 枚，平均 20.2 枚，花冠直径 3.8cm。在辽宁兴城，果实 9 月下旬成熟，单果重 195g，纵径 7.5cm，横径 7.1cm，倒卵圆形；果皮绿色或黄绿色；果心小，5 心室；果肉白色，肉质中粗，松脆，汁液多，味酸甜或甜；含可溶性固形物 12.33%，可滴定酸 0.15%；品质中上等。

2. Chazili

Origin and Distribution Chazili, originated in Tianbaoshan area, Pingyi County, Shandong Province, is grown mainly in Pingyi, Feixian, and Tengxian, etc.

Main Characters Tree: moderately vigorous, productive, medium precocity. Leaf: 10.4cm × 7.7cm in size, ovate. Initial leaf: green tinged with red. Flower: white bud tinged with light pink on edge, 5 to 7 flowers per cluster in average of 6.0; stamen number: 18 to 22, averaging 20.2; corolla diameter: 3.8cm. Fruit: matures in late September in Xingcheng, Liaoning Province, 195g per fruit, 7.5cm long, 7.1cm wide, obovate, green or yellowish-green skin, small core, 5 locules, flesh white, mid-coarse, crisp tender, juicy, sour-sweet or sweet; TSS 12.33%, TA 0.15%; quality above medium.

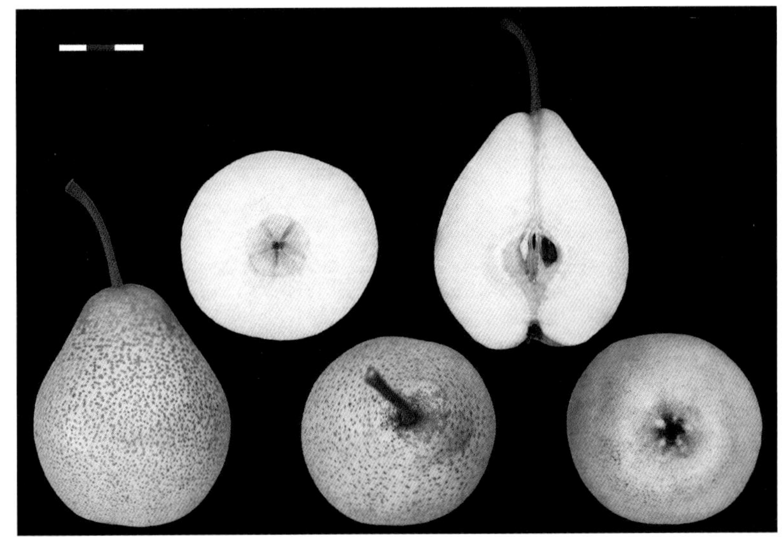

3. 茌梨

来源及分布　2n=34，又名慈梨，原产地在山东省茌平、牟平和莱阳一带。在山东莱阳、栖霞栽培最多，河北、江苏、辽西、陕西、新疆南部等地亦有分布。

主要性状　树势强。始果年龄中或晚，丰产。叶片长 12.3cm，宽 7.1cm，广卵圆形，初展叶暗红色。花蕾白色，边缘粉红色，每花序 4 ~ 5 朵花，平均 4.3 朵；雄蕊 19 ~ 21 枚，平均 20.0 枚；花冠直径 3.8cm。在辽宁兴城，果实 9 月下旬或 10 月上旬成熟，果实多不整齐，梗端多向一侧弯曲，卵圆形或纺锤形；单果重 233g，纵径 8.5cm，横径 7.4cm；果皮黄绿色，果点大而明显；果心中大，5 心室，果肉浅黄白色，肉质细，松脆，汁液多，味浓甜，石细胞小而少；含可溶性固形物 14.15%，可滴定酸 0.19%；品质上等。

3. Chili

Origin and Distribution　Chili (2n=34), also known as Cili, was originated in Chiping, Muping, or Laiyang of Shandong Province, is grown mainly in Laiyang and Qixia, also grown in Hebei, Jiangsu, western part of Liaoning, Shaanxi, and southern Xingjiang, etc.

Main Characters　Tree: vigorous, productive, medium or late precocity. Leaf: 12.3cm × 7.1cm in size, broadly ovate. Initial leaf: dark red. Flower: white bud tinged with pink on edge, 4 to 5 flowerers per cluster in average of 4.3; stamen number: 19 to 21, averaging 20.0; corolla diameter: 3.8cm. Fruit: matures in late September or early October in Xingcheng, Liaoning Province, ovate or spindle-shaped, irregular, usually bend to one side at the stem base, 233g per fruit, 8.5cm long, 7.4cm wide, yellowish-green skin with dots large and obvious, medium core, 5 locules, flesh pale yellowish-white, fine, crisp tender, juicy, very sweet, with less and very small grit cells; TSS 14.15%, TA 0.19%; quality good.

4. 崇化大梨

来源及分布 2n=34，别名水冬瓜、大泡梨，金川雪梨自然实生，原产四川省金川县安宁乡，为当地优良品种之一。

主要性状 树势强，丰产，始果年龄早。叶片长12.3cm，宽9.3cm，椭圆形，初展叶绿色，着红色，有茸毛。花蕾白色，边缘粉红色，每花序5～6朵花，平均5.3朵；雄蕊19～21枚，平均20.0枚；花冠直径4.4cm。在辽宁兴城，果实9月下旬成熟。单果重253g，纵径9.3cm，横径7.4cm，葫芦形或纺锤形；果皮绿黄色；果心中大，5或4心室；果肉白色，肉质中粗，松脆，汁液多，味淡甜；含可溶性固形物11.23%，可滴定酸0.11%；品质中上等。

4. Chonghua Dali

Origin and Distribution Chonghua Dali (2n=34), also known as Shuidonggua, or Dapaoli, selected from the seedlings of Jinchuan Xueli in Anning village, Jinchuan County, Sichuan Province, is one of the fine varieties there.

Main Characters Tree: vigorous, productive; early precocity. Leaf: 12.3cm × 9.3cm in size, elliptical. Initial leaf: green tinged with red, pubescent. Flower: white bud tinged with pink on edge, 5 to 6 flowers per cluster in average of 5.3; stamen number: 19 to 21, averaging 20.0; corolla diameter: 4.4cm. Fruit: matures in late September in Xingcheng, Liaoning Province, pyriform or spindle-shaped, 253g per fruit, 9.3cm long, 7.4cm wide, greenish-yellow skin, medium core, 5 or 4 locules, flesh white, mid-coarse, crisp tender, juicy, light sweet; TSS 11.23%, TA 0.11%; quality above medium.

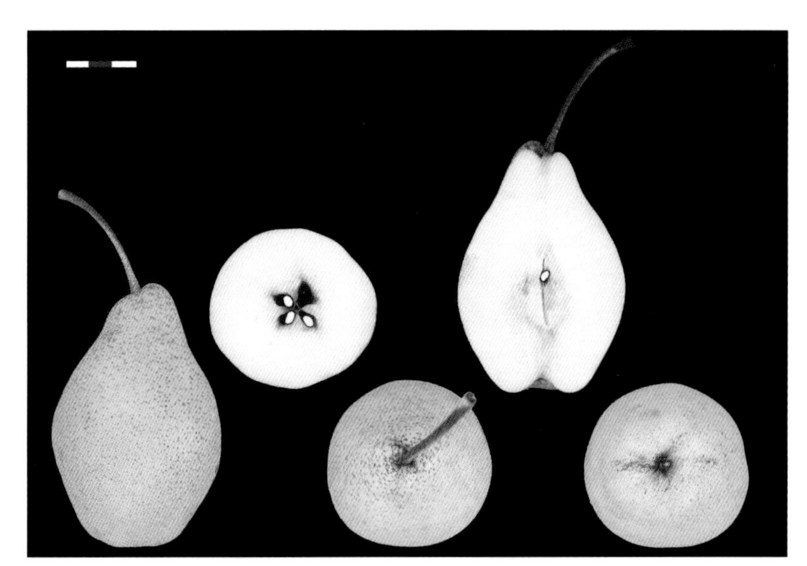

5. 砀山酥梨

来源及分布 2n=34，又名酥梨、砀山梨，原产安徽省砀山县，是古老的地方优良品种。在辽宁、山西、山东、陕西、河南、四川、云南、新疆等地有大面积栽培。

主要性状 树势强，始果年龄中等，丰产。叶片长10.3cm，宽8.9cm，卵圆形，初展叶红色，微显绿色。花蕾白色，每花序5～7朵花，平均6.2朵；雄蕊19～25枚，平均20.7枚；花冠直径4.2cm。在辽宁兴城，果实9月下旬成熟，单果重239g，纵径7.6cm，横径7.7cm，圆柱形，萼端平截稍宽；果皮绿黄色，贮后黄色，果肩部或有小锈块；果心小，5心室，果心周围有大颗粒石细胞；果肉白色，较细或中粗，肉质疏松，汁液多，味甜；含可溶性固形物12.45%，可滴定酸0.10%；品质上等。耐贮藏。

5. Dangshan Suli

Origin and Distribution Dangshan Suli (2n=34), also known as Suli, Dangshanli, originated in Dangshan County, Anhui Province, is one of the best old pear varieties in China which is widely grown in Liaoning, Shanxi, Shandong, Shaanxi, Henan, Sichuan, Yunnan, and Xinjiang, etc.

Main Characters Tree: vigorous, productive, medium precocity. Leaf: 10.3cm × 8.9cm in size, ovate. Initial leaf: red with less green. Flower: white bud, 5 to 7 flowers per cluster in average of 6.2; stamen number: 19 to 25, averaging 20.7; corolla diameter: 4.2cm. Fruit: matures in late September in Xingcheng, Liaoning Province, cylinder-shaped, calyx basin wide, 239g per fruit, 7.6cm long, 7.7cm wide, greenish-yellow skin, turning to yellow after storage, small core, 5 locules, flesh white, relatively fine or mid-coarse, with more large stone cells around core, tender, juicy, sweet; TSS 12.45%, TA 0.10%; quality good; storage life long.

6. 大水核子

来源及分布 2n=3x=51，原产江苏省，主要分布在江苏淮北地区。

主要性状 树势强，产量中等，始果年龄中等。叶片长12.4cm，宽9.4cm，阔卵圆形，初展叶红色。花蕾白色，边缘淡粉红色，每花序6～7朵花，平均6.5朵；雄蕊20～30枚，平均25.5枚；花冠直径4.2cm。在辽宁兴城，果实9月中旬成熟，单果重291g，纵径8.4cm，横径8.1cm，阔倒卵圆形或阔倒圆锥形；果皮绿色或黄绿色；果心小，5心室；果肉白色，肉质细，松脆，汁液特多，味甜酸；含可溶性固形物10.20%，可滴定酸0.33%；品质中上等。

6. Dashuihezi

Origin and Distribution Dashuihezi (2n=3x=51), originated in Jiangsu Province, is grown mainly in Huaibei area of Jiangsu Province.

Main Characters Tree: vigorous, mid-productive, medium precocity. Leaf: 12.4cm × 9.4cm in size, broadly ovate. Initial leaf: red. Flower: white bud tinged with light pink on edge, 6 to 7 flowers per cluster in average of 6.5; stamen number: 20 to 30, averaging 25.5; corolla diameter: 4.2cm. Fruit: matures in mid-September in Xingcheng, Liaoning Province, 291g per fruit, 8.4cm long, 8.1cm wide, broadly obovate or turbination, green or yellowish-green skin, small core, 5 locules, flesh white, fine, crisp tender, vey juicy, sweet-sour; TSS 10.20%, TA 0.33%; quality above medium.

7. 大窝窝梨

来源及分布 2n=34，又名大凹凹梨，原产山东省，在山东青岛等地有栽培。

主要性状 树势强，丰产，始果年龄早或中等，植株寿命长。叶片长11.5cm，宽6.9cm，卵圆形，初展叶红色，微显绿色。花蕾白色，边缘粉红色，每花序4～6朵花，平均4.6朵；雄蕊19～25枚，平均21.3枚；花冠直径4.1cm。在辽宁兴城，果实9月下旬成熟，单果重255g，纵径7.9cm，横径7.6cm，近圆形；果皮绿色或黄绿色；果心中大，5心室；果肉白色，肉质中粗，松脆，汁液多，味酸甜；含可溶性固形物11.45%，可滴定酸0.25%；品质中上等。果实耐贮藏。

7. Dawowoli

Origin and Distribution Dawowoli (2n=34), also known as Daaoaoli, originated in Shandong Province, is grown mainly in Qingdao, etc.

Main Characters Tree: vigorous, productive, early or medium precocity, long-lived. Leaf: 11.5cm × 6.9cm in size, ovate. Initial leaf: red with less green. Flower: white bud tinged with pink on edge, 4 to 6 flowers per cluster in average of 4.6; stamen number: 19 to 25, averaging 21.3; corolla diameter: 4.1cm. Fruit: matures in late September in Xingcheng, Liaoning Province, 255g per fruit, 7.9cm long, 7.6cm wide, sub-round, green or yellowish-green skin, medium core, 5 locules, flesh white, mid-coarse, crisp tender, juicy, sour-sweet; TSS 11.45%, TA 0.25%; quality above medium; storage life long.

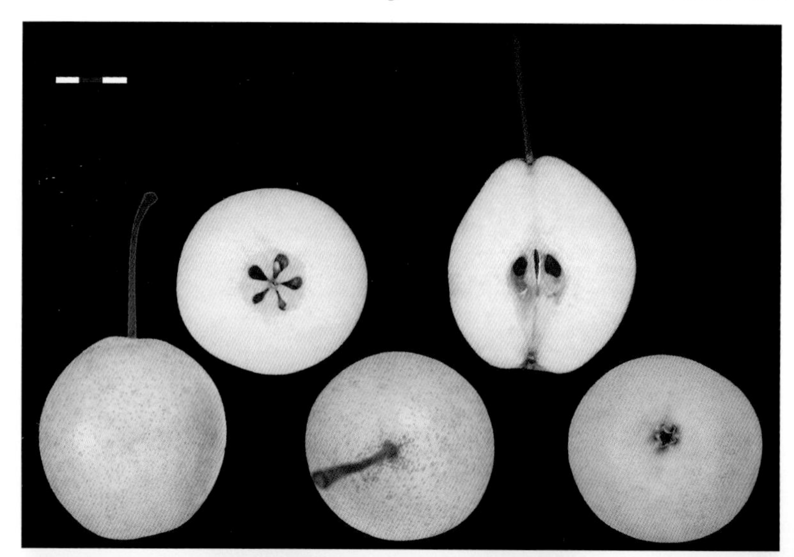

8. 冬果梨

来源及分布 2n=34，原产甘肃省黄河流域，在甘肃、宁夏等地有栽培。

主要性状 树势强，始果年龄中等，丰产。叶片长 11.2cm，宽 6.9cm，卵圆形，初展叶红色，有茸毛。花蕾白色，小花蕾边缘淡粉红色，每花序 5 ~ 7 朵花，平均 5.8 朵；雄蕊 22 ~ 29 枚，平均 24.7 枚；花冠直径 5.4cm。在辽宁兴城，果实 9 月下旬成熟，单果重 250g，纵径 8.5cm，横径 7.6cm，果实倒卵圆形；果皮绿黄色，贮藏后黄色，果心中等大，5 心室，果肉白色，肉质稍粗，松脆或稍紧，汁液多，味酸甜；含可溶性固形物 12.23%，可滴定酸 0.23%；品质中或中上等。果实耐贮藏。

8. Dongguoli

Origin and Distribution Dongguoli (2n=34), originated in Gansu Province along Yellow River basin, is grown mainly in Gansu and Ningxia, etc.

Main Characters Tree: vigorous, productive, medium precocity. Leaf: 11.2cm × 6.9cm in size, ovate. Initial leaf: red, pubescent. Flower: white bud, small one tinged with light pink on edge, 5 to 7 flowers per cluster in average of 5.8; stamen number: 22 to 29, averaging 24.7; corolla diameter: 5.4cm. Fruit: matures in late September in Xingcheng, Liaoning Province, 250g per fruit, 8.5cm long, 7.6cm wide, obovate, greenish-yellow skin, turning to yellow after storage, medium core, 5 locules, flesh white, somewhat coarse, crisp tender or somewhat dense, juicy, sour-sweet; TSS 12.23%, TA 0.23%; medium or above in quality; storage life long.

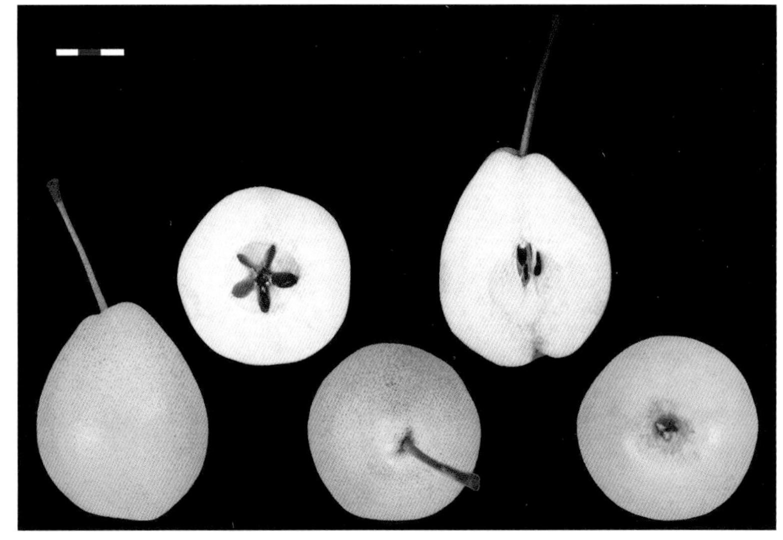

9. 冬黄梨

来源及分布 2n=34，原产新疆，在新疆霍城、伊宁等地有栽培。

主要性状 树势强，丰产，始果年龄中等。叶片长10.0cm，宽7.3cm，卵圆形或阔卵圆形，初展叶红色，微显绿色。花蕾白色，边缘粉红色，花瓣边缘亦粉红色，每花序7～9朵花，平均7.6朵；雄蕊20～23枚，平均20.9枚；花冠直径3.8cm。在辽宁兴城，果实9月下旬成熟，单果重86g，纵径5.5cm，横径5.4cm，卵圆形；果皮绿黄色，部分果实阳面有条状暗红晕；果心中大，3～5心室；果肉淡黄白色，肉质较细，松脆，汁液多，味甜或淡甜；含可溶性固形物12.95%，可滴定酸0.096%；品质中上等。果实耐贮藏。

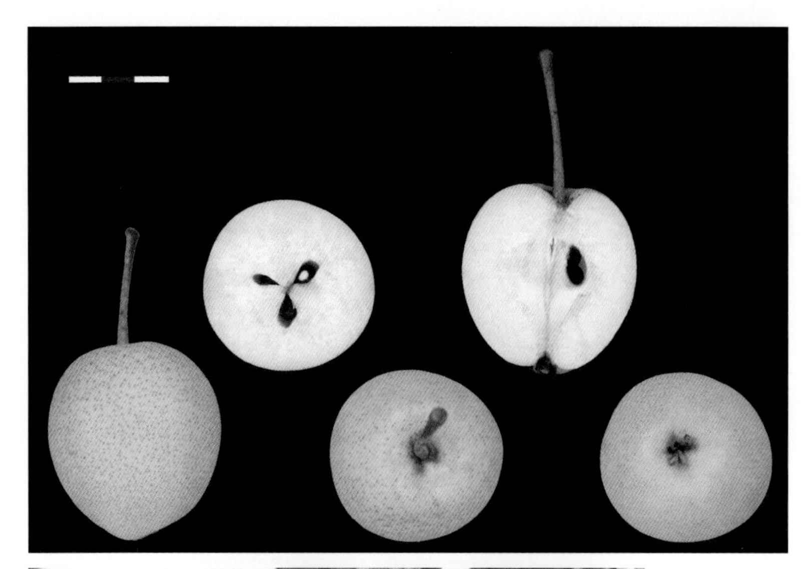

9. Donghuangli

Origin and Distribution Donghuangli (2n=34), originated in Xinjiang, is grown mainly in Huocheng and Yining, etc.

Main Characters Tree: vigorous, productive, medium precocity. Leaf: 10.0cm × 7.3cm in size, ovate or broadly ovate. Initial leaf: red with less green. Flower: white bud tinged with pink on edge, edge of the petals also light pink, 7 to 9 flowers per cluster in average of 7.6; stamen number: 20 to 23, averaging 20.9; corolla diameter: 3.8cm. Fruit: matures in late September in Xingcheng, Liaoning Province, 86g per fruit, 5.5cm long, 5.4cm wide, ovate, greenish-yellow skin, some covered with striped dark red on the side exposed to the sun, medium core, 3 to 5 locules, flesh pale yellowish-white, relatively fine, crisp tender, juicy, sweet or light sweet; TSS 12.95%, TA 0.096%; quality above medium; storage life long.

10. 鹅黄

来源及分布 原产安徽省，在砀山地区栽培较多。

主要性状 树势强，丰产，始果年龄晚。叶片长 11.1cm，宽 6.5cm，卵圆形，初展叶红色，微显绿色。花蕾白色，边缘粉红色，每花序 4 ~ 7 朵花，平均 5.4 朵；雄蕊 20 ~ 28 枚，平均 24.7 枚；花冠直径 4.2cm。在辽宁兴城，果实 9 月下旬成熟，单果重 118g，纵径 6.1cm，横径 6.0cm，近圆形或椭圆形；果皮绿黄色或黄色；果心小，5 心室；果肉白色，肉质中粗，疏松，稍脆，汁液多，味甜；含可溶性固形物 13.00%，可滴定酸 0.19%；品质中上等。

10. Ehuang

Origin and Distribution Ehuang, originated in Anhui Province, is grown mainly in Dangshan County.

Main Characters Tree: vigorous, productive, late precocity. Leaf: 11.1cm × 6.5cm in size, ovate. Initial leaf: red with less green. Flower: white bud tinged with pink on edge, 4 to 7 flowers per cluster in average of 5.4; stamen number: 20 to 28, averaging 24.7; corolla diameter: 4.2cm. Fruit: matures in late September in Xingcheng, Liaoning Province, 118g per fruit, 6.1cm long, 6.0cm wide, sub-round or elliptical, greenish-yellow or yellow skin, small core, 5 locules, flesh white, mid-coarse, tender, somewhat crisp, juicy, sweet; TSS 13.00%, TA 0.19%; quality above medium.

11. 粉红宵

来源及分布 2n=34，原产辽宁省，在辽宁西部和河北昌黎等地有栽培。

主要性状 树势强，产量中等，始果年龄中或晚。叶片长10.5cm，宽6.6cm，卵圆形，初展叶红色，微显绿色。花蕾白色，或边缘粉红色，每花序6～8朵花，平均6.9朵；雄蕊19～20枚，平均19.5枚；花冠直径4.3cm。在辽宁兴城，果实9月下旬成熟，单果重166g，纵径6.3cm，横径6.6cm，圆柱形或近圆形；果皮绿黄色，有淡红晕；果心中大，4或5心室；果肉白色，较细，疏松，稍脆，汁液多，味甜；含可溶性固形物13.20%，可滴定酸0.18%；品质中上等。果实耐贮藏。

11. Fenhongxiao

Origin and Distribution Fenhongxiao (2n=34), originated in Liaoning Province, is grown mainly in Western Liaoning, also in Changli County of Hebei Province.

Main Characters Tree: vigorous, medium production, medium or late precocity. Leaf: 10.5cm × 6.6cm in size, ovate. Initial leaf: red with less green. Flower: white bud, or tinged with pink on edge, 6 to 8 flowers per cluster in average of 6.9; stamen number: 19 to 20, averaging 19.5; corolla diameter: 4.3cm. Fruit: matures in late September in Xingcheng, Liaoing Province, 166g per fruit, 6.3cm long, 6.6cm wide, cylindrical or sub-round, greenish-yellow skin tinged with red blush, medium core, 4 or 5 locules, flesh white, fine, tender, somewhat crisp, juicy, sweet; TSS 13.20%, TA 0.18%; quality above medium; storage life long.

12. 凤县鸡腿梨

来源及分布 2n=34，又名凤县龙口梨，原产陕西省，在凤县、略阳等地有栽培。

主要性状 树势强，丰产。叶片长12.8cm，宽8.6cm，卵圆形，初展叶红色，微显绿色。花蕾白色，每花序5～6朵花，平均5.8朵；雄蕊19～23枚，平均20.8枚；花冠直径4.2cm。在辽宁兴城，果实9月中、下旬成熟，单果重210g，纵径7.4cm，横径7.1cm，阔倒卵圆形；果皮绿黄色或黄色；果心中大，5心室；果肉白色，肉质细，松脆，汁液多，味淡甜；含可溶性固形物11.63%，可滴定酸0.19%；品质中上等。

12. Fengxian Jituili

Origin and Distribution Fengxian Jituili (2n=34), also known as Fengxian Longkouli, originated in Shaanxi Province, is grown mainly in Fengxian, Lueyang County, etc.

Main Characters Tree: vigorous, productive. Leaf: 12.8cm × 8.6cm in size, ovate. Initial leaf: red with less green. Flower: white bud, 5 to 6 flowers per cluster in average of 5.8; stamen number: 19 to 23, averaging 20.8; corolla diameter: 4.2cm. Fruit: matures in mid or late September in Xingcheng, Liaoning Province, 210g per fruit, 7.4cm long, 7.1cm wide, broadly obovate, greenish-yellow or yellow skin, medium core, 5 locules, flesh white, fine, crisp tender, juicy, light sweet; TSS 11.63%, TA 0.19%; quality above medium.

13. 佛见喜

来源及分布 2n=34，原产河北省，在遵化、蓟县、迁安、兴隆等地有栽培。

主要性状 树势强，丰产，始果年龄晚。叶片长8.0cm，宽6.1cm，卵圆形，初展叶红色。花蕾白色，边缘浅粉红色，每花序6～7朵花，平均6.6朵；雄蕊19～24枚，平均21.1枚；花冠直径4.6cm。在辽宁兴城，果实9月中旬成熟，单果重132g，纵径5.4cm，横径6.4cm，扁圆形；果皮绿黄色或黄白色，有鲜红晕；果心中大，4或5心室；果肉白色，肉质细，松脆，汁液多，味甜；含可溶性固形物13.68%，可滴定酸0.31%；品质中上等。

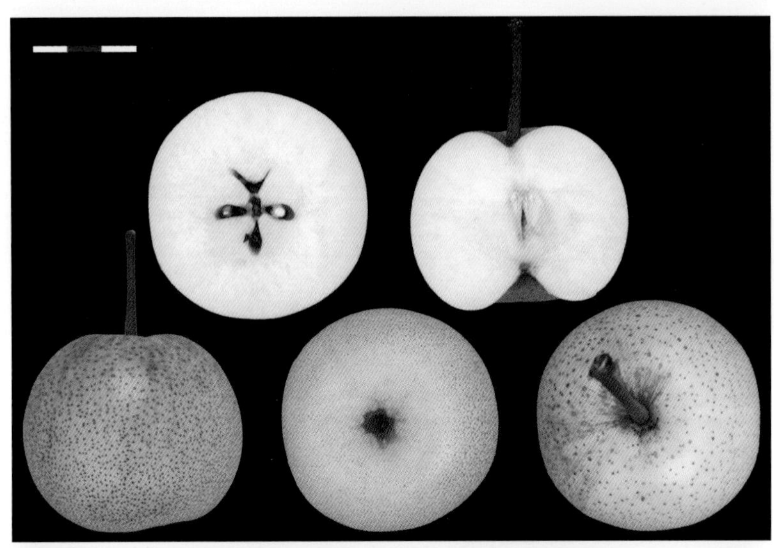

13. Fojianxi

Origin and Distribution Fojianxi (2n=34), originated in Hebei Province, is grown mainly in Zunhua, Jixian, Qian'an, and Xinglong County, etc.

Main Characters Tree: vigorous, productive, late precocity. Leaf: 8.0cm × 6.1cm in size, ovate. Initial leaf: red. Flower: white bud tinged with light pink on edge, 6 to 7 flowers per cluster in average of 6.6; stamen number: 19 to 24, averaging 21.1; corolla diameter: 4.6cm. Fruit: matures in mid-September in Xingcheng, Liaoning Province, 132g per fruit, 5.4cm long, 6.4cm wide, oblate, greenish-yellow or whitish-yellow skin covered with bright red blush, medium core, 4 or 5 locules, flesh white, fine, crisp tender, juicy, sweet; TSS 13.68%, TA 0.31%; quality above medium.

14. 海城慈梨

来源及分布 2n=34，原产辽宁省，在海城等地有少量栽培。

主要性状 树势中庸，较丰产，始果年龄中等。叶片长11.7cm，宽7.5cm，卵圆形，初展叶红色。花蕾白色，边缘粉红色，每花序4～6朵花，平均5.5朵；雄蕊20～27枚，平均23.1枚；花冠直径3.9cm。在辽宁兴城，果实9月中、下旬成熟，单果重301g，纵径8.9cm，横径8.3cm，圆锥形，不规则；果皮绿黄色或黄色；果心中大或小，多数5心室，少数6心室；果肉白色，肉质中粗，脆，稍紧密，汁液中多，味淡甜或甜；含可溶性固形物12.68%，可滴定酸0.22%；品质中上等。

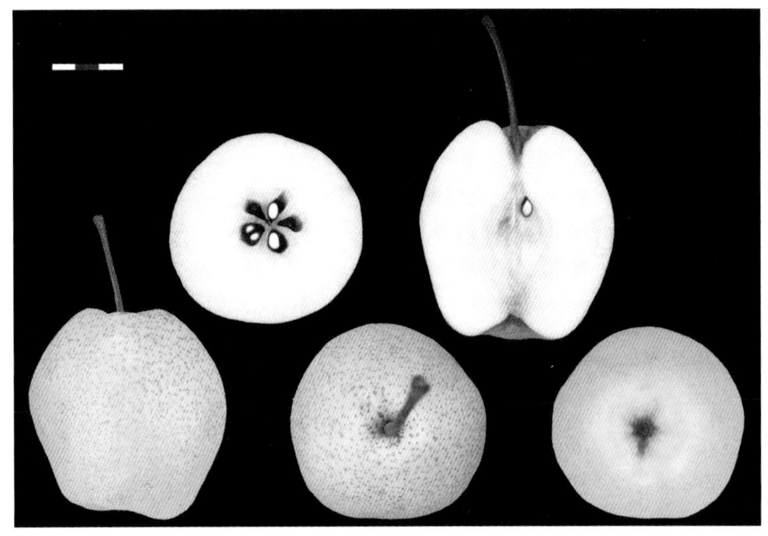

14. Haicheng Cili

Origin and Distribution Haicheng Cili (2n=34), originated in Liaoning Province, is grown in Haicheng limitedly.

Main Characters Tree: moderately vigorous, relatively productive, medium precocity. Leaf: 11.7cm × 7.5cm in size, ovate. Initial leaf: red. Flower: white bud tinged with pink on edge, 4 to 6 flowers per cluster in average of 5.5; stamen number: 20 to 27, averaging 23.1; corolla diameter: 3.9cm. Fruit: matures in mid or late September in Xingcheng, Liaoning Province, 301g per fruit, 8.9cm long, 8.3cm wide, conical, irregular, greenish-yellow or yellow skin, medium or small core, 5 locules, less 6, flesh white, mid-coarse, crisp, somewhat dense, mid-juicy, sweet or light sweet; TSS 12.68%, TA 0.22%; quality above medium.

15. 汉源白梨

来源及分布　2n=34，原产四川省，在汉源等地有栽培。

主要性状　树势强，丰产，始果年龄晚。叶片长10.0cm，宽8.0cm，卵圆形，初展叶绿色，微着红色。花蕾白色，每花序4～7朵花，平均5.5朵；雄蕊20～21枚，平均20.1枚；花冠直径3.4cm。在辽宁兴城，果实9月下旬成熟，单果重118g，纵径6.4cm，横径5.9cm，阔倒卵圆形；果皮黄绿色；果心大或中大，5心室；果肉白色，肉质中粗，疏松，汁液多，味甜；含可溶性固形物11.20%，可滴定酸0.06%；品质中或中上等。

15. Hanyuan Baili

Origin and Distribution Hanyuan Baili (2n=34), originated in Sichuan Province, is grown mainly in Hanyuan, etc.

Main Characters Tree: vigorous, productive, late precocity. Leaf: 10.0cm × 8.0cm in size, ovate. Initial leaf: green slightly tinged with red. Flower: white bud, 4 to 7 flowers per cluster in average of 5.5; stamen number: 20 to 21, averaging 20.1; corolla diameter: 3.4cm. Fruit: matures in late September in Xingcheng, Liaoning Province, 118g per fruit, 6.4cm long, 5.9cm wide, broadly obovate, yellowish-green skin, large or medium core, 5 locules, flesh white, mid-coarse, tender, juicy, sweet; TSS 11.20%, TA 0.06%; medium or above in quality.

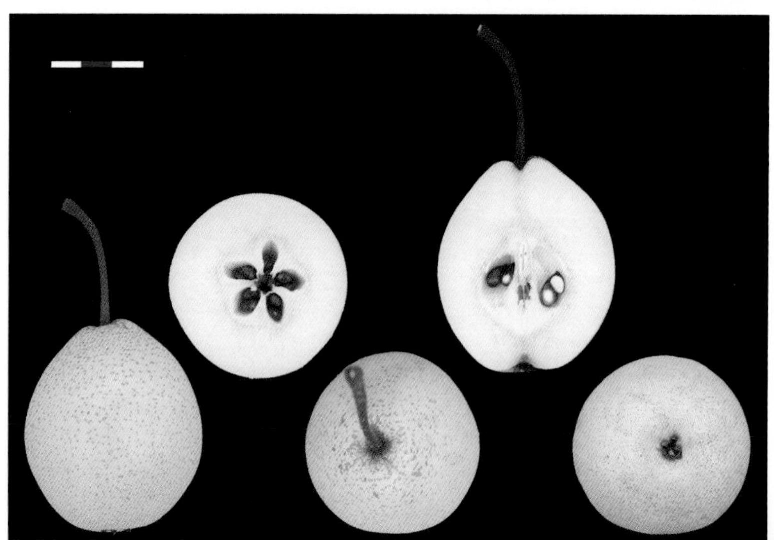

16. 红枝母秧

来源及分布 2n=34，原产河北省兴隆县，系红梨种子实生繁殖而来。

主要性状 树势强，丰产，始果年龄晚。叶片长 10.9cm，宽 7.2cm，卵圆形，初展叶红色，微显绿色。花蕾白色，边缘淡粉红色，每花序 3 ~ 8 朵花，平均 7.1 朵；雄蕊 20 ~ 24 枚，平均 21.2 枚；花冠直径 3.8cm。在辽宁兴城，果实 9 月中、下旬成熟，单果重 95g，纵径 5.0cm，横径 5.7cm，扁圆形或圆形；果皮绿黄色或黄白色，阳面有鲜红晕；果心中大，5 心室；果肉白色，肉质细，松脆，汁液中多或多，味甜或酸甜；含可溶性固形物 13.18%，可滴定酸 0.35%；品质中上等果实耐贮藏。

16. Hongzhi Muyang

Origin and Distribution Hongzhi Muyang (2n=34), a chance seedling of Hongli, originated in Xinglong County, Hebei Province.

Main Characters Tree: vigorous, productive, late precocity. Leaf: 10.9cm × 7.2cm in size, ovate. Initial leaf: red with less green. Flower: white bud tinged with light pink on edge, 3 to 8 flowers per cluster in average of 7.1; stamen number: 20 to 24, averaging 21.2; corolla diameter: 3.8cm. Fruit: matures in mid or late September in Xingcheng, Liaoning Province, 95g per fruit, 5.0cm long, 5.7cm wide, oblate or round, greenish-yellow or whitish-yellow skin covered with bright red on the side exposed to thc sun, medium core, 5 locules, flesh white, fine, crisp tender, juicy or mid-juicy, sweet or sour-sweet; TSS 13.18%, TA 0.35%; quality above medium; storage life long.

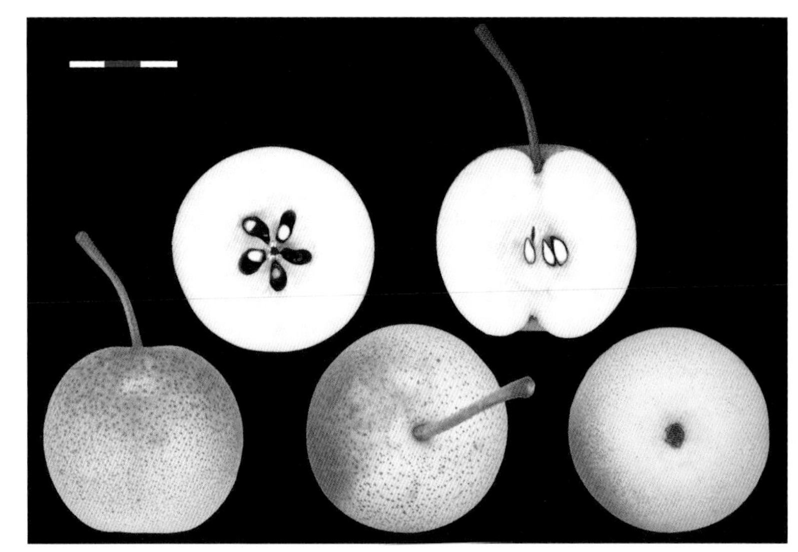

17. 黄县长把

来源及分布 2n=34，又名大把梨、天生梨，原产山东省黄县，自然实生，在当地栽培较多。

主要性状 树势强，丰产。叶片长 12.3cm，宽 8.2cm，卵圆形，初展叶红色，叶脉绿色。花蕾白色，边缘粉红色，每花序 5 ~ 8 朵花，平均 6.3 朵；雄蕊 19 ~ 26 枚，平均 21.0 枚；花冠直径 4.0cm。在辽宁兴城，果实 9 月下旬成熟，单果重 110g，纵径 6.3cm，横径 5.9cm，阔倒卵圆形或椭圆形；果皮绿色，贮藏后转黄色；果心中等大，5 心室；果肉白色，肉质松脆，较细，汁液多，味甜酸；含可溶性固形物 13.41%，可滴定酸 0.33%；品质中上等。果实耐贮藏。

17. Huangxian Changba

Origin and Distribution Huangxian Changba (2n=34), also known as Dabali, or Tianshengli, a chance seedling, was originated in Huangxian, Shandong Province, is grown mainly there.

Main Characters Tree: vigorous, productive. Leaf: 12.3cm × 8.2cm in size, ovate. Initial leaf: red with vein green. Flower: white bud tinged with pink on edge, 5 to 8 flowers per cluster in average of 6.3; stamen number: 19 to 26, averaging 21.0; corolla diameter: 4.0cm. Fruit: matures in late September in Xingcheng, Liaoning Province, 110g per fruit, 6.3cm long, 5.9cm wide, broadly obovate or oblong, green skin, turning to yellow after storage, medium core, 5 locules, flesh white, crisp tender, relatively fine, juicy, sweet-sour; TSS 13.41%, TA 0.33%; quality above medium; storage life long.

18. 金川雪梨

来源及分布 2n=34，别名大金鸡腿梨，原产四川省金川县，大金川两岸分布较多。

主要性状 树势强，丰产，始果年龄中等。叶片长13.1cm，宽8.3cm，卵圆形，初展叶绿色，微着红色。花蕾白色，边缘粉红色，每花序4～7朵花，平均5.9朵；雄蕊18～23枚，平均20.0枚；花冠直径4.0cm。在辽宁兴城，果实9月下旬成熟，单果重221g，纵径8.7cm，横径7.2cm，葫芦形或倒卵圆形；果皮绿黄色；果心中大，5心室；果肉白色，肉质中粗，松脆，汁液多，味淡甜或酸甜；含可溶性固形物11.27%，可滴定酸0.20%；品质中上等。

18. Jinchuan Xueli

Origin and Distribution Jinchuan Xueli (2n=34), also known as Dajin Jituili, was originated in Jinchuan County, Sichuan Province, is grown mainly along Big Jinchuan River.

Main Characters Tree: vigorous, productive, medium precocity. Leaf: 13.1cm × 8.3cm in size, ovate. Initial leaf: green, slightly tinged with red. Flower: white bud tinged with pink on edge, 4 to 7 flowers per cluster in average of 5.9; stamen number: 18 to 23, averaging 20.0; corolla diameter: 4.0cm. Fruit: matures in late September in Xingcheng, Liaoning Province, 221g per fruit, 8.7cm long, 7.2cm wide, pyriform or obovate, greenish-yellow skin, medium core, 5 locules, flesh white, mid-coarse, crisp tender, juicy, light sweet or sour sweet; TSS 11.27%, TA 0.20%; quality above medium.

19. 金锤子

来源及分布 原产辽宁省。

主要性状 树势强，丰产。叶片长9.9cm，宽5.8cm，卵圆形，初展叶红色，微显绿色。花蕾白色，边缘浅粉红色，每花序6～9朵花，平均7.5朵；雄蕊19～21枚，平均20.0枚；花冠直径3.9cm。在辽宁兴城，果实9月中、下旬成熟，单果重165g，纵径8.1cm，横径6.2cm，纺锤形，萼端平截，似锤子状；果皮绿黄色；果心中大，5心室；果肉白色，肉质中粗，松脆，汁液多，味甜酸；含可溶性固形物10.87%，可滴定酸0.23%；品质中上等。果实耐贮藏。

19. Jinchuizi

Origin and Distribution Jinchuizi, was originated in Liaoning Province.

Main Characters Tree: vigorous, productive. Leaf: 9.9cm × 5.8cm in size, ovate. Initial leaf: red with less green. Flower: white bud tinged with light pink on edge, 6 to 9 flowers per cluster in average of 7.5; stamen number: 19 to 21, averaging20.0; corolla diameter: 3.9cm. Fruit: matures in mid or late September in Xingcheng, Liaoning Province, 165g per fruit, 8.1cm long, 6.2cm wide, spindle-shaped, flat at basin, like a hammer, greenish-yellow skin, medium core, 5 locules, flesh white, mid-coarse, crisp tender, juicy, sweet-sour; TSS 10.87%, TA 0.23%; quality above medium; storage life long.

20. 金花梨

来源及分布　2n=34，又名林檎梨，原产四川省金川县，在四川金川及云南昆明等地有栽培。

主要性状　树势强，丰产，始果年龄中等。叶片长 11.8cm，宽 8.2cm，卵圆形，初展叶绿色，着红色。花蕾白色，边缘粉红色，每花序 5 ～ 6 朵花，平均 5.7 朵；雄蕊 20 ～ 24 枚，平均 21.3 枚；花冠直径 4.5cm。在辽宁兴城，果实 9 月下旬成熟，单果重 291g，纵径 9.2cm，横径 7.7cm，长圆形或倒卵圆形；果皮绿黄色；果心小，5 心室；果肉白色，肉质细，松脆，汁液多，味甜；含可溶性固形物 12.78%，可滴定酸 0.14%；品质上等。果实耐贮藏。

20. Jinhuali

Origin and Distribution Jinhuali (2n=34), also known as Linqinli, originated in Jinchuan County, Sichuan Province, is grown mainly in Jinchuan, Sichuan Province, also in Kunming, Yunnan Province.

Main Characters Tree: vigorous, productive, medium precocity. Leaf: 11.8cm × 8.2cm in size, ovate. Initial leaf: green tinged with red. Flower: white bud tinged with pink on edge, 5 to 6 flowers per cluster in average of 5.7; stamen number: 20 to 24, averaging 21.3; corolla diameter: 4.5cm. Fruit: matures in late September in Xingcheng, Liaoning Province, 291g per fruit, 9.2cm long, 7.7cm wide, long round or obovate, greenish-yellow skin, small core, 5 locules, flesh white, fine, crisp tender, juicy, sweet; TSS 12.78%, TA 0.14%; quality good; storage life long.

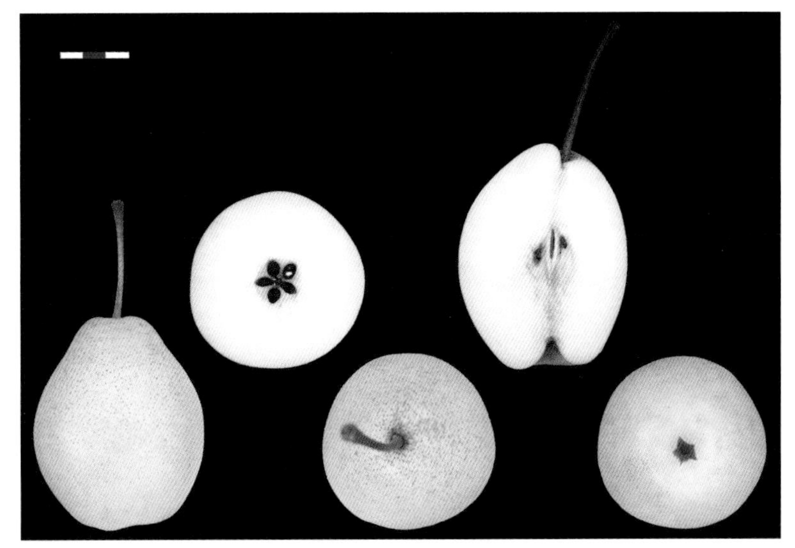

21. 金梨

来源及分布　2n=34，原产山西省，在万荣、隰县、蒲县等地有栽培。

主要性状　树势强，丰产，始果年龄早或中等。叶片长 10.6cm，宽 6.7cm，卵圆形，初展叶红色，微显绿色。花蕾白色，边缘粉红色，每花序 4 ～ 6 朵花，平均 4.9 朵；雄蕊 21 ～ 31 枚，平均 25.2 枚；花冠直径 3.9cm。在辽宁兴城，果实 9 月下旬成熟，单果重 390g，纵径 9.6cm，横径 8.6cm，圆锥形或长圆形；果皮绿色或绿黄色，贮藏后黄色；果心中等大，5 心室；果肉白色，肉质脆，中粗，汁液多，味淡甜或酸甜；含可溶性固形物 11.33%，可滴定酸 0.25%；品质中上等。果实耐贮藏。

21. Jinli

Origin and Distribution Jinli (2n=34), originated in Shanxi Province, is grown mainly in Wanrong, Xixian, and Puxian, etc.

Main Characters Tree: vigorous, productive; early or medium precocity. Leaf: 10.6cm × 6.7cm in size, ovate. Initial leaf: red with less green. Flower: white bud tinged with pink on edge, 4 to 6 flowers per cluster in average of 4.9; stamen number: 21 to 31, averaging 25.2; corolla diameter: 3.9cm. Fruit: matures in late September in Xingcheng, Liaoning Province, 390g per fruit, 9.6cm long, 8.6cm wide, conical or oblong, green or greenish-yellow skin, turning to yellow after storage, medium core, 5 locules, flesh white, crisp, mid-coarse, juicy, light sweet or sour-sweet; TSS 11.33%, TA 0.25%; quality above medium; storage life long.

22. 连云港黄梨

来源及分布 2n=34，原产江苏省连云港市，为当地优良品种之一。

主要性状 树势强，丰产。叶片长 9.1cm，宽 9.1cm，阔卵圆形或圆形，初展叶红色，微显绿色，有茸毛。花蕾白色，边缘浅粉红色，每花序 5 ～ 7 朵花，平均 6.1 朵；雄蕊 20 ～ 25 枚，平均 21.3 枚；花冠直径 4.6cm。在辽宁兴城，果实 9 月下旬成熟，单果重 145g，纵径 7.4cm，横径 6.1cm，长圆形；果皮绿黄色；果心中大，4 或 5 心室；果肉白色，肉质中粗，紧密而脆，汁液多，味淡甜；含可溶性固形物 9.34%；品质中上等。

22. Lianyungang Huangli

Origin and Distribution Lianyungang Huangli (2n=34), originated in Lianyungang, Jiangsu Province, is one of good pear varieties there.

Main Characters Tree: vigorous, productive. Leaf: 9.1cm × 9.1cm in size, broadly ovate or round. Initial leaf: red with less green, pubescent. Flower: white bud tinged with light pink on edge, 5 to 7 flowers per cluster in average of 6.1; stamen number: 20 to 25, averaging 21.3; corolla diameter: 4.6cm. Fruit: matures in late September in Xingcheng, Liaoning Province, 145g per fruit, 7.4cm long, 6.1cm wide, long round, greenish-yellow skin, medium core, 4 or 5 locules, flesh white, mid-coarse, crisp, somewhat dense, juicy, light sweet; TSS 9.34%; quality above medium.

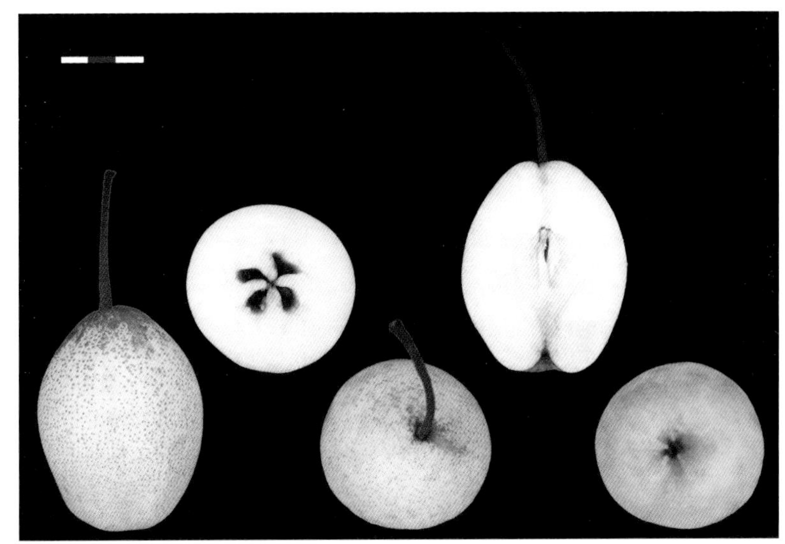

23. 懋功梨

来源及分布 2n=34，原产四川省小金县，在小金县及泸定县等地有栽培。

主要性状 树势强，较丰产，始果年龄晚。叶片长13.7cm，宽9.6cm，阔卵圆形，初展叶红色，微显绿色。花蕾白色，边缘粉红色，每花序5～7朵花，平均6.7朵；雄蕊23～27枚，平均25.8枚；花冠直径3.7cm。在辽宁兴城，果实9月上旬成熟，单果重151g，纵径7.3cm，横径6.3cm，卵圆形或圆锥形；果皮绿黄色或黄白色；果心小或中大，5心室；果肉白色，肉质松脆，细，汁液多，味甜或淡甜；含可溶性固形物13.50%，可滴定酸0.07%；品质中上或上等。

23. Maogongli

Origin and Distribution Maogongli (2n=34), originated in Xiaojin County, Sichuan Province, is grown mainly in Xiaojin and Luding , etc.

Main Characters Tree: vigorous, relatively productive, late precocity. Leaf: 13.7cm × 9.6cm in size, broadly ovate. Initial leaf: red with less green. Flower: white bud tinged with pink on edge, 5 to 7 flowers per cluster in average of 6.7; stamen number: 23 to 27, averaging 25.8; corolla diameter: 3.7cm. Fruit: matures in early September in Xingcheng, Liaoning Province, 151g per fruit, 7.3cm long, 6.3cm wide, ovate or conical, greenish-yellow or whitish-yellow skin, small or medium core, 5 locules; flesh white, crisp tender, fine, juicy, sweet or light sweet; TSS 13.50%, TA 0.07%; good or above medium in quality.

24. 蜜梨

来源及分布 2n=34，原产河北省，在昌黎、青龙、兴隆、迁安、蓟县等地有栽培。

主要性状 树势强，丰产，始果年龄较晚。叶片长 9.4cm，宽 6.7cm，卵圆形，初展叶绿色，着红色，有茸毛。花蕾白色，边缘淡粉红色，每花序 4 ~ 7 朵花，平均 5.6 朵；雄蕊 20 ~ 22 枚，平均 20.2 枚，花冠直径 3.8cm。在辽宁兴城，果实 9 月下旬成熟，单果重 93g，纵径 5.3cm，横径 5.5cm，长圆形或圆锥形；果皮绿黄色，阳面微有红晕；果心中大，4 或 5 心室；果肉白色，肉质细，松脆，汁液多，味甜；含可溶性固形物 11.93%，可滴定酸 0.18%；品质中上等。果实耐贮藏。

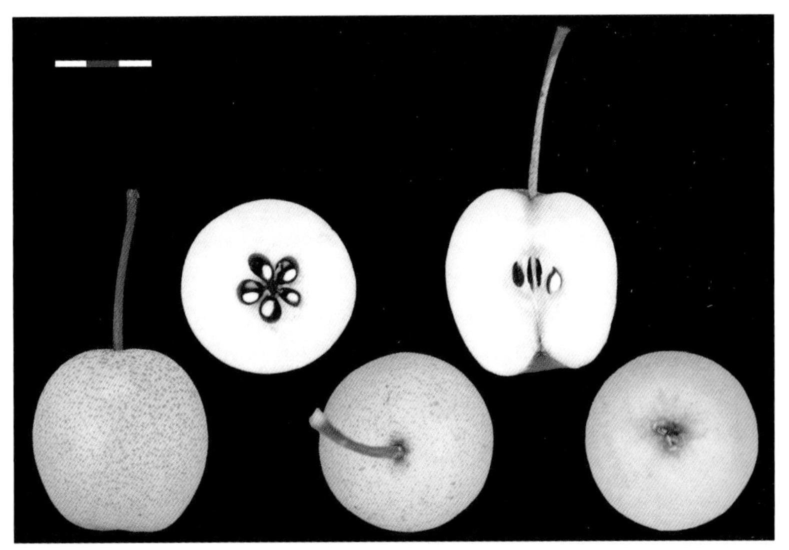

24. Mili

Origin and Distribution Mili (2n=34), originated in Hebei Province, is cultivated in Changli, Qinglong, Xinglong, Qian'an, and Jixian, etc.

Main Characters Tree: vigorous, productive; late precocity. Leaf: 9.4cm × 6.7cm in size, ovate. Initial leaf: green tinged with red, pubescent. Flower: white bud tinged with light pink on edge, 4 to 7 flowers per cluster in average of 5.6; stamen number: 20 to 22, averaging 20.2; corolla diameter: 3.8cm. Fruit: matures in late September in Xingcheng, Liaoning Province, 93g per fruit, 5.3cm long, 5.5cm wide, long round or conical, greenish-yellow skin tinged with light red on the side exposed to the sun, medium core, 4 or 5 locules, flesh white, fine, crisp tender, juicy, sweet; TSS 11.93%, TA 0.18%; quality above medium; storage life long.

25. 苹果梨

来源及分布 2n=34，来源不详，在吉林省龙井、和龙、延吉等地有大面积栽培，在辽宁、甘肃及内蒙古和新疆等地亦栽培较多。

主要性状 树势中庸，丰产，始果年龄中等。叶片长12.0cm，宽6.4cm，卵圆形，初展叶绿色，微显红色，有茸毛。花蕾白色，边缘粉红色，每花序8 ~ 12朵花，平均9.2朵；雄蕊20 ~ 27枚，平均21.5枚；花冠直径4.1cm。在辽宁兴城，果实9月下旬或10月上旬成熟，单果重212g，纵径6.4cm，横径7.7cm，扁圆形，不规整，形态似苹果；果皮绿黄色，阳面有红晕；果心极小，5心室，果肉白色，肉质细，脆，汁液多，味酸甜；含可溶性固形物12.77%，可滴定酸0.26%；品质上等。果实耐贮藏。

25. Pingguoli

Origin and Distribution Pingguoli (2n=34), origin unclear, is grown mainly in Longjing, Helong, and Yanji, etc. It is also grown commercially in Liaoning, Gansu, Inner Mongolia, and Xinjiang, etc.

Main Characters Tree: moderately vigorous, productive; medium precocity. Leaf: 12.0cm × 6.4cm in sizc, ovate. Initial leaf: green with less red, pubescent. Flower: white bud tinged with pink on edge, 8 to 12 flowers per cluster in average of 9.2; stamen number: 20 to 27, averaging 21.5; corolla diameter: 4.1cm. Fruit: matures in late September or early October in Xingcheng, Liaoning Province, 212g per fruit, 6.4cm long, 7.7cm wide, oblate, irregular, shape like apple, greenish-yellow skin covered with red blush on the side exposed to the sun, extremely small core, 5 locules, flesh white, fine, crisp, juicy, sour-sweet; TSS 12.77%, TA 0.26%; quality good; storage life long.

26. 棋盘香梨

来源及分布 2n=34，原产新疆叶城，在南疆有栽培。

主要性状 树势强，丰产，始果年龄中或晚。叶片长 7.4cm，宽 5.8cm，阔卵圆形，初展叶绿色，着红色。花蕾白色，每花序 5 ～ 7 朵花，平均 6.0 朵；雄蕊 20 ～ 24 枚，平均 20.7 枚；花冠直径 3.3cm。在辽宁兴城，果实 9 月下旬成熟，单果重 107g，纵径 5.4cm，横径 6.0cm，近圆形。果皮绿黄色，阳面有淡的红晕；果心大，5 心室；果肉白色，肉质中粗，疏松，汁液中多，味酸甜或甜；含可溶性固形物 13.18%，可滴定酸 0.46%；品质中等或中上等。

26. Qipan Xiangli

Origin and Distribution Qipan Xiangli (2n=34), originated in Yecheng, Xinjiang, is grown mainly in southern Xinjiang.

Main Characters Tree: vigorous, productive, precocity late or medium. Leaf: 7.4cm × 5.8cm in size, broadly ovate. Initial leaf: green tinged with red. Flower: white bud, 5 to 7 flowers per cluster in average of 6.0; stamen number: 20 to 24, averaging 20.7; corolla diameter: 3.3cm. Fruit: matures in late September in Xingcheng, Liaoning Province, 107g per fruit, 5.4cm long, 6.0cm wide, sub-round; greenish-yellow skin covered with light red on the side exposed to the sun, large core, 5 locules; flesh white, mid-coarse, mid-juicy, tender, sour-sweet or sweet; TSS 13.18%, TA 0.46%; quality medium or above.

27. 青龙甜

来源及分布　2n=34，原产河北省青龙县，当地叫野梨。

主要性状　树势强，丰产，始果年龄中等。叶片长8.3cm，宽6.9cm，卵圆形，初展叶绿色，着红色。花蕾白色，边缘粉红色，每花序5～6朵花，平均5.7朵；雄蕊20～23枚，平均20.9枚；花冠直径4.2cm。在辽宁兴城，果实9月中、下旬成熟，单果重180g，纵径7.1cm，横径6.7cm，倒卵圆形；果皮绿黄色或黄色；果心中大，5心室；果肉白色，肉质中粗，松脆，汁液多，味甜；含可溶性固形物12.70%，可滴定酸0.16%；品质中上等。

27. Qinglongtian

Origin and Distribution Qinglongtian (2n=34), called Yeli there, was originated in Qinglong County, Hebei Province.

Main Characters Tree: vigorous, productive, medium precocity. Leaf: 8.3cm × 6.9cm in size, ovate. Initial leaf: green tinged with red. Flower: white bud tinged with pink on edge, 5 to 6 flowers per cluster in average of 5.7; stamen number: 20 to 23, averaging 20.9; corolla diameter: 4.2cm. Fruit: matures in mid or late September in Xingcheng, Liaoning Province, 180g per fruit, 7.1cm long, 6.7cm wide, obovate, greenish-yellow or yellow skin, medium core, 5 locules, flesh white, mid-coarse, crisp tender, juicy, sweet; TSS 12.70%, TA 0.16%; quality above medium.

28. 青皮蜂蜜

来源及分布 2n=34，原产四川省泸定县，为自然实生。

主要性状 树势强，丰产。叶片长 12.0cm，宽 8.4cm，阔卵圆形，初展叶红色，微显绿色。花蕾白色，边缘淡粉红色，每花序 4 ～ 6 朵花，平均 5.3 朵；雄蕊 20 ～ 23 枚，平均 20.6 枚；花冠直径 3.8cm。在辽宁兴城，果实 9 月下旬成熟，单果重 142g，纵径 6.3cm，横径 6.5cm，阔倒卵圆形；果皮绿色或黄绿色；果心中大，4 或 5 心室；果肉白色，肉质中粗，汁液多，松脆，味甜；含可溶性固形物 11.90%，可滴定酸 0.11%；品质中等或中上等。

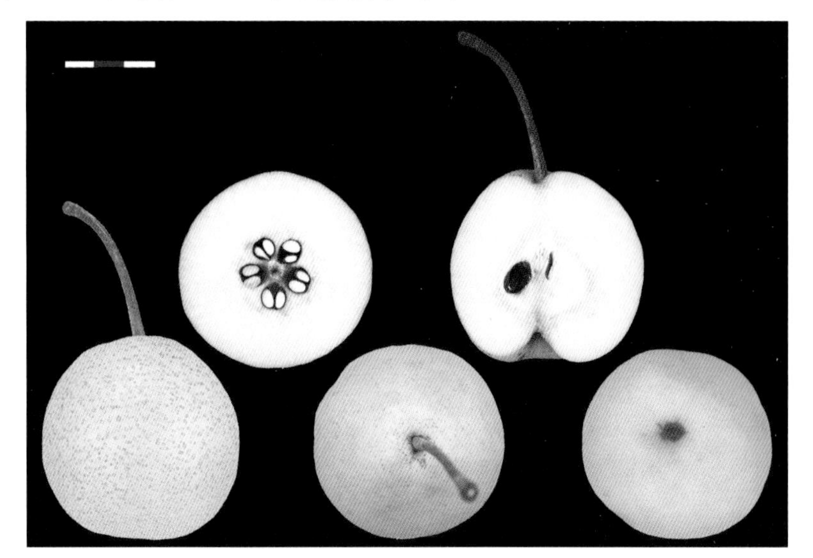

28. Qingpi Fengmi

Origin and Distribution Qingpi Fengmi (2n=34), a chance seedling, was originated in Luding County, Sichuan Province.

Main Characters Tree: vigorous, productive. Leaf: 12.0cm × 8.4cm in size, broadly ovate. Initial leaf: red with less green. Flower: white bud tinged with light pink on edge, 4 to 6 flowers per cluster in average of 5.3; stamen number: 20 to 23, averaging 20.6; corolla diameter: 3.8cm. Fruit: matures in late September in Xingcheng, Liaoning Province, 142g per fruit, 6.3cm long, 6.5cm wide, broadly obovate, green or yellowish-green skin, medium core, 4 or 5 locules, flesh white, mid-coarse, juicy, crisp tender, sweet; TSS 11.90%, TA 0.11%; medium or above in quality.

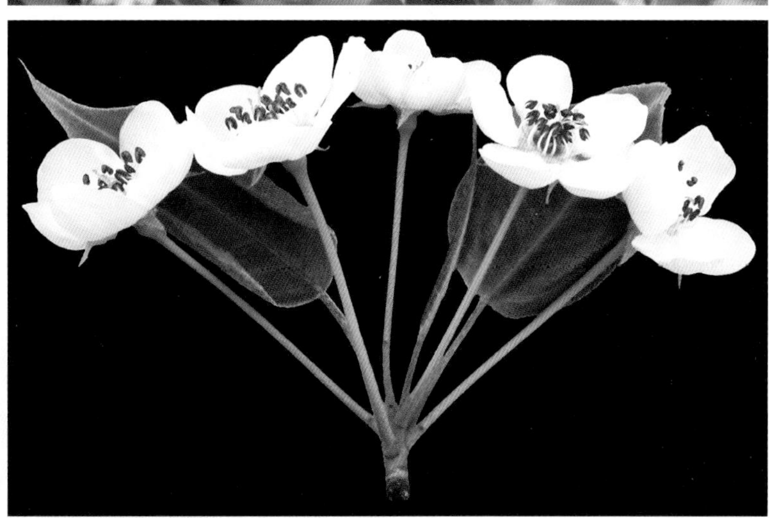

29. 秋白梨

来源及分布 2n=34，原产河北省，在河北燕山山脉和辽宁西部地区有大面积栽培，辽宁鞍山地区亦有较多栽培。

主要性状 树势中庸，产量中等，始果年龄中等。叶片长 10.1cm，宽 6.5cm，卵圆形，初展叶绿色，微显红色。花蕾白色，边缘粉红色，每花序 5 ～ 7 朵花，平均 6.1 朵；雄蕊 20 ～ 23 枚，平均 20.8 枚；花冠直径 3.7cm。在辽宁兴城，果实 9 月下旬至 10 月上旬成熟，单果重 150g，纵径 6.5cm，横径 6.5cm，近圆形或长圆形；果皮绿黄色，贮后为黄色；果心小，心室数 3 ～ 5；果肉白色，肉质细，脆，紧密，汁液较多，味甜；含可溶性固形物 13.50%，可滴定酸 0.21%；品质上等。果实耐贮藏。

29. Qiubaili

Origin and Distribution Qiubaili (2n=34), originated in Hebei Province, is widely cultivated in Yanshan mountains in Hebei Province, also in Anshan and western part of Liaoning Province.

Main Characters Tree: moderately vigorous, medium productive, medium precocity. Leaf : 10.1cm × 6.5cm in size, ovate. Initial leaf: green with less red. Flower: white bud tinged with pink on edge, 5 to 7 flowers per cluster, averaging 6.1; stamen number: 20 to 23 in average of 20.8; corolla diameter: 3.7cm. Fruit: matures in late September or early October in Xingcheng, Liaoning Province, 150g per fruit, 6.5cm long, 6.5cm wide, sub-round or long round, greenish-yellow skin, turning to yellow after storage, small core, 3 to 5 locules, flesh white, fine, crisp and dense, juicy, sweet; TSS 13.50%, TA 0.21%; quality good; storage life long.

30. 栖霞大香水

来源及分布 2n=34，又名南宫茌，原产山东省栖霞市，为茌梨较好的授粉品种，在陕西渭北地区也有发展。

主要性状 树势中庸，产量中等或较高，始果年龄早或中等。叶片长10.6～14.1cm，宽6.3～9.8cm，卵圆形或长卵圆形，初展叶红色，带极少的绿色。花蕾白色，边缘粉红色，每花序4～6朵花，平均4.9朵；雄蕊19～20枚，平均19.8枚；花冠直径3.9cm。在辽宁兴城，果实9月下旬成熟，单果重101g，纵径6.0cm，横径5.6cm，长圆形；果皮采收时绿色，贮后转黄绿色或黄色；果心中大，5心室；果肉白色，肉质松脆，较细，汁液多，味酸甜；含可溶性固形物11.05%，可滴定酸0.25%；品质中上或上等。果实耐贮藏。

30. Qixia Daxiangshui

Origin and Distribution Qixia Daxiangshui, also known as Nangongchi, originated in Qixia, Shandong Province, is one of good pollination varieties of Chi Li. It is also cultivated in the north of Weihe River in Shaanxi Province.

Main Characters Tree: moderately vigorous, medium or high production, early or medium precocity. Leaf: 10.6-14.1cm × 6.3-9.8cm in size, ovate or long ovate. Initial leaf: red with very less green. Flower: white bud tinged with pink on edge, 4 to 6 flowers per cluster in average of 4.9; stamen number: 19 to 20, averaging 19.8; corolla diameter: 3.9cm. Fruit: matures in late September in Xingcheng, Liaoning Province, 101g per fruit, 6.0cm long, 5.6cm wide, long round, green skin when harvested, turning to greenish-yellow or yellow after storage, medium core, 5 locules, flesh white, crisp tender, relatively fine, juicy, sour-sweet; TSS 11.05%, TA 0.25%; above medium or good in quality; storage life long.

31. 水白梨

来源及分布 原产河北省青龙县。

主要性状 树势强，丰产，始果年龄中等。叶片长 10.5cm，宽 6.9cm，卵圆形，初展叶绿色，微显红色，有茸毛。花蕾白色，边缘浅粉红色，每花序 5 ~ 8 朵花，平均 6.2 朵；雄蕊 20 枚；花冠直径 3.9cm。在辽宁兴城，果实 9 月下旬成熟，单果重 183g，纵径 6.2cm，横径 7.0cm，扁圆形或近圆形；果皮黄绿色，部分果实阳面着淡红晕；果心小，4 或 5 心室；果肉白色，肉质细，松脆，汁液多，味酸甜；含可溶性固形物 13.10%，可滴定酸 0.33%；品质中上等或上等。

31. Shuibaili

Origin and Distribution Shuibaili, was originated in Qinglong County, Hebei Province.

Main Characters Tree: vigorous, productive, medium precocity. Leaf: 10.5cm × 6.9cm in size, ovate. Initial leaf: green with less red, pubescent. Flower: white bud tinged with light pink on edge, 5 to 8 flowers per cluster in average of 6.2; stamen number: 20; corolla diameter: 3.9cm. Fruit: matures in late September in Xingcheng, Liaoning Province, 183g per fruit, 6.2cm long, 7.0cm wide, oblate or sub-round, yellowish-green skin, some covered with light red on the side exposed to the sun, small core, 4 or 5 locules, flesh white, fine, crisp tender, juicy, sour-sweet; TSS 13.10%, TA 0.33%; good or above medium in quality.

32. 水红宵

来源及分布 2n=34，产于辽宁绥中、北镇及河北青龙等地，为红梨的品系之一。

主要性状 树势强，较丰产，始果年龄中或晚。叶片长 11.2cm，宽 8.0cm，卵圆形，初展叶绿色，着红色，有茸毛。花蕾白色，边缘粉红色，每花序 4 ～ 7 朵花，平均 5.9 朵；雄蕊 19 ～ 20 枚，平均 19.9 枚；花冠直径 3.8cm。在辽宁兴城，果实 9 月下旬或 10 月上旬成熟，单果重 176g，纵径 6.3cm，横径 6.8cm，短圆柱形；果皮绿黄色，阳面有鲜红晕；果心小，3 ～ 5 心室；果肉白色，肉质细，松脆，汁液多，味酸甜；含可溶性固形物 11.67%，可滴定酸 0.16%；品质上等。

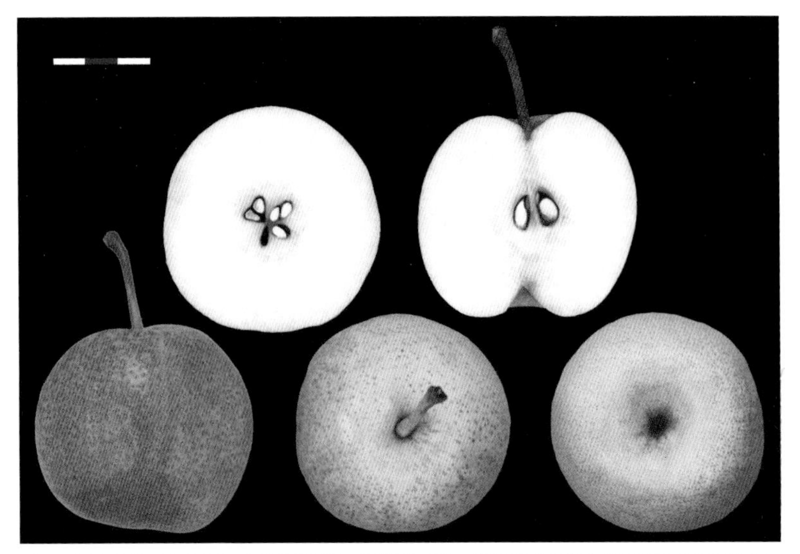

32. Shuihongxiao

Origin and Distribution Shuihongxiao (2n=34), one of Hongli strains, is cultivated in Suizhong, Beizhen of Liaoning Province, also in Qinglong County of Hebei Province.

Main Characters Tree: vigorous, productive, medium or late precocity. Leaf: 11.2cm × 8.0cm in size, ovate. Initial leaf: green tinged with red, pubescent. Flower: white bud tinged with pink on edge, 4 to 7 flowers per cluster in average of 5.9; stamen number: 19 to 20, averaging 19.9; corolla diameter: 3.8cm. Fruit: matures in late September or early October in Xingcheng, Liaoning Province, 176g per fruit, 6.3cm long, 6.8cm wide, short cylindrical, greenish-yellow skin covered with bright red on the side exposed to the sun, small core, 3 to 5 locules, flesh white, fine, crisp tender, juicy, sour-sweet; TSS 11.67%, TA 0.16%; quality good.

33. 水葫芦梨

来源及分布 原产安徽省砀山县。

主要性状 树势中庸。叶片长 10.4cm，宽 6.4cm，卵圆形，初展叶绿色，着红色。花蕾白色，边缘粉红色，每花序 3 ~ 5 朵花，平均 3.9 朵；雄蕊 16 ~ 20 枚，平均 19.1 枚；花冠直径 4.2cm。在辽宁兴城，果实 9 月中旬成熟，单果重 130.7g，纵径 6.1cm，横径 6.2cm，圆形或长圆形；果皮绿黄色，梗洼有条锈；果心中大，5 心室；果肉淡黄白色，肉质中粗或细，松脆，汁液多，味甜酸；含可溶性固形物 11.47%，可滴定酸 0.38%；品质中上等。

33. Shuihululi

Origin and Distribution Shuihululi, was originated in Dangshan County, Anhui Province.

Main Characters Tree: moderately vigorous. Leaf: 10.4cm × 6.4cm in size, ovate. Initial leaf: green tinged with red. Flower: white bud tinged with pink on edge, 3 to 5 flowers per cluster in average of 3.9; stamen number: 16 to 20, averaging 19.1; corolla diameter: 4.2cm. Fruit: matures in mid-September in Xingcheng, Liaoning Province, 130.7g per fruit, 6.1cm long, 6.2cm wide, round or long round, greenish-yellow skin with striped russet on stalk cavity, medium core, 5 locules, flesh pale yellowish-white, fine or mid-coarse, crisp tender, juicy, sweet-sour; TSS 11.47%, TA 0.38%; quality above medium.

34. 绥中谢花甜

来源及分布 2n=34，原产辽宁省绥中县，在当地有少量栽培。

主要性状 树势强，较丰产，始果年龄中或晚；叶片长 10.1cm，宽 6.6cm，卵圆形，初展叶红色。花蕾白色，每花序 7 ～ 8 朵花，平均 7.5 朵；雄蕊 20 枚；花冠直径 4.3cm。在辽宁兴城，果实 9 月中、下旬成熟，单果重 135g，纵径 6.0cm，横径 6.6cm，圆柱形、近圆形或扁圆形；果皮黄白色，阳面有淡红晕或鲜红晕；果心中大，4 或 5 心室；果肉白色，肉质较细，松脆，汁液多，味甜；含可溶性固形物 13.28%，可滴定酸 0.22%；品质中上等。果实耐贮藏。

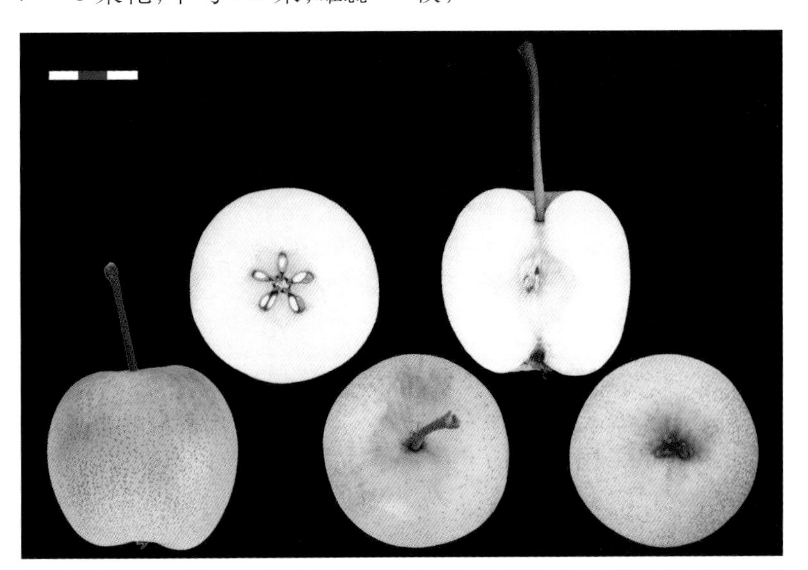

34. Suizhong Xiehuatian

Origin and Distribution Suizhong Xiehuatian (2n=34), originated in Suizhong County, Liaoning Province, is grown there limitedly.

Main Characters Tree: vigorous, productive, medium or late precocity. Leaf: 10.1cm × 6.6cm in size, ovate. Initial leaf: red. Flower: white bud, 7 to 8 flowers per cluster in average of 7.5; stamen number: 20; corolla diameter: 4.3cm. Fruit: matures in mid or late September in Xingcheng, Liaoning Province, 135g per fruit, 6.0cm long, 6.6cm wide, cylindrical, sub-round, or oblate, yellowish-white skin covered with light or bright red on the side exposed to the sun, medium core , 4 or 5 locules, flesh white, fine, crisp tender, juicy, sweet; TSS 13.28%, TA 0.22%; quality above medium; storage life long.

35. 酥木梨

来源及分布 2n=34，原产甘肃省，在甘肃陇中及河西一带有栽培。

主要性状 树势中庸，较丰产，始果年龄中等。叶片长 11.4cm，宽 7.7cm，卵圆形，初展叶绿色，有茸毛。花蕾白色，边缘淡粉红色，每花序 6 ~ 9 朵花，平均 7.2 朵；雄蕊 20 ~ 24 枚，平均 21.5 枚；花冠直径 4.4cm。在辽宁兴城，果实 9 月中旬成熟，单果重 124g，纵径 5.7cm，横径 6.0cm，近圆形；果皮绿黄色或黄白色，阳面有淡红晕；果心大，5 心室；果肉淡黄白色，肉质粗，松软，汁液中多，味甜酸，稍涩，含可溶性固形物 12.23%，可滴定酸 0.50%；品质中等或中上等。果实不耐贮藏。

35. Sumuli

Origin and Distribution Sumuli (2n=34), originated in Gansu Province, is grown mainly in Longzhong and Hexi corridor in Gansu Province.

Main Characters Tree: vigorous, productive, medium precocity. Leaf: 11.4cm × 7.7cm in size, ovate. Initial leaf: green, pubescent. Flower: white bud tinged with light pink on edge, 6 to 9 flowers per cluster in average of 7.2; stamen number: 20 to 24, averaging 21.5; corolla diameter: 4.4cm. Fruit: matures in mid-September in Xingcheng, Liaoning Province, 124g per fruit, 5.7cm long, 6.0cm wide, sub-round, greenish-yellow or yellowish-white skin covered with light red blush on the side exposed to the sun, large core, 5 locules, flesh pale yellowish-white, coarse, tender and soft, mid-juicy, sweet-sour, a little astringent; TSS 12.23%, TA 0.50%; quality medium or above; storage life short.

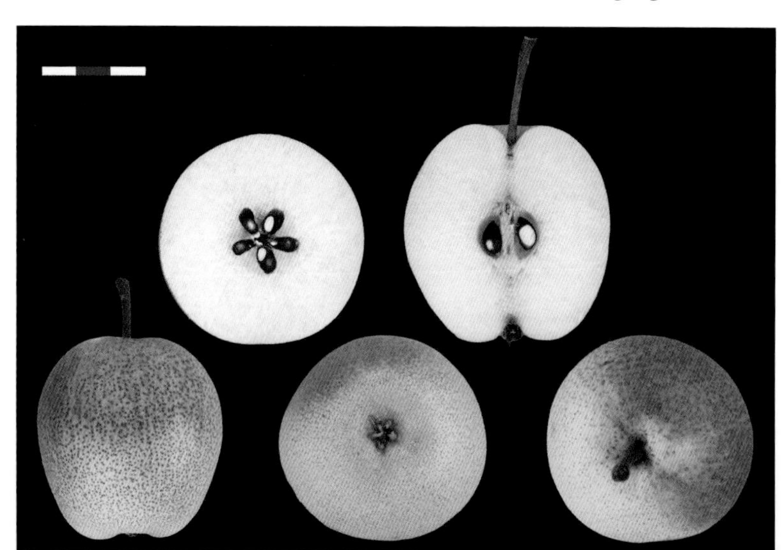

36. 胎黄梨

来源及分布 2n=34，原产河北省交河县。

主要性状 树势中庸，丰产，连续结果能力差。叶片长9.9cm，宽8.0cm，宽卵圆形，初展叶绿色，微显红色，有茸毛。花蕾白色，每花序7～8朵花，平均7.4朵；雄蕊19～22枚，平均20.1枚；花冠直径4.0cm。在辽宁兴城，果实9月下旬成熟，单果重65.3g，纵径5.0cm，横径4.7cm，长圆形；果皮绿黄色，有浅的条红；果心中大，3～5心室；果肉白色，肉质中粗，松脆，汁液多，味甜，含可溶性固形物13.12%，可滴定酸0.14%；品质中上或上等。

36. Taihuangli

Origin and Distribution Taihuangli (2n=34), was originated in Jiaohe County, Hebei Province.

Main Characters Tree: moderately vigorous, productive, alternate bearing. Leaf: 9.9cm × 8.0cm in size, broadly ovate. Initial leaf: green with less red, pubescent. Flower: white bud, 7 to 8 flowers per cluster in average of 7.4; stamen number: 19 to 22, averaging 20.1; corolla diameter: 4.0cm. Fruit: matures in late September in Xingcheng, Liaoning Province, 65.3g per fruit, 5.0cm long, 4.7cm wide, long round, greenish-yellow skin tinged with light striped red, medium core, 3 to 5 locules, flesh white, mid- coarse, crisp tender, juicy, sweet; TSS 13.12%, TA 0.14%; above medium or good in quality.

37. 小花梨

来源及分布　2n=34，原产江苏省，在连云港有少量栽培。

主要性状　树势强，丰产，始果年龄中或晚。叶片长 9.3cm，宽 7.1cm，近圆形或阔卵圆形；初展叶红色，带有绿色。花蕾白色，边缘粉红色，每花序 4 ～ 6 朵花，平均 5.0 朵；雄蕊 22 ～ 31 枚，平均 26.1 枚；花冠直径 3.5cm。在辽宁兴城，果实 9 月中旬成熟，单果重 144g，纵径 6.8cm，横径 6.2cm，长圆形或椭圆形；果皮黄绿色；果心中大，5 或 6 心室；果肉白色，肉质中粗，松脆，汁液多，味淡甜；含可溶性固形物 11.20%，可滴定酸 0.30%；品质中上等。

37. Xiaohuali

Origin and Distribution　Xiaohuali (2n=34), originated in Jiangsu Province, is grown in Lianyungang limitedly.

Main Characters　Tree: vigorous, productive, medium or late precocity. Leaf: 9.3cm × 7.1cm in size, sub-round or broadly ovate. Initial leaf: red with green. Flower: white bud tinged with pink on edge, 4 to 6 flowers per cluster in average of 5.0; stamen number: 22 to 31, averaging 26.1; corolla diameter: 3.5cm. Fruit: matures in mid-September in Xingcheng, Liaoning Province, 144g per fruit, 6.8cm long, 6.2cm wide, long round or oblong, yellowish-green skin, medium core, 5 or 6 locules, flesh white, mid-coarse, crisp tender, juicy, light sweet; TSS 11.20%, TA 0.30%; quality above medium.

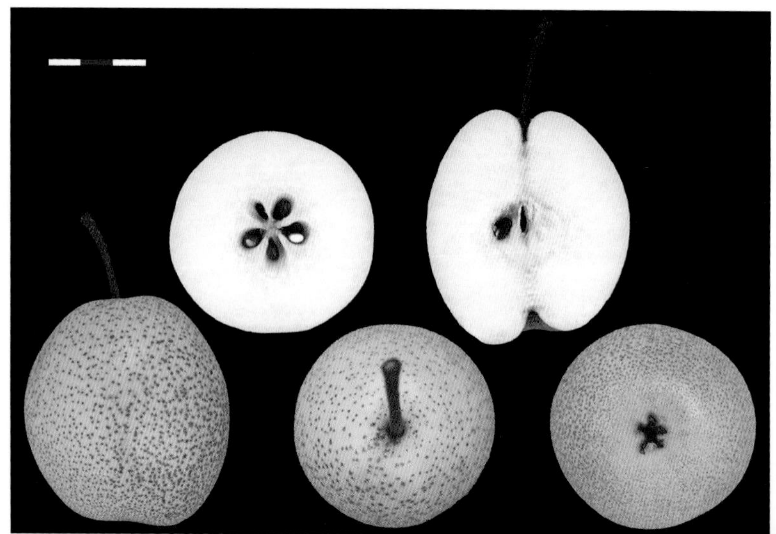

38. 夏梨

来源及分布 2n=34，原产山西省，在原平、五台、榆次等地有栽培。

主要性状 树势强，丰产，始果年龄晚，树龄长。叶片长11.2cm，宽6.9cm，卵圆形，初展叶红色，带有绿色。花蕾白色，每花序6～8朵花，平均6.8朵；雄蕊19～21枚，平均20.0枚；花冠直径4.2cm。在辽宁兴城，果实9月下旬成熟，单果重146g，纵径6.3cm，横径6.4cm，倒卵圆形；果皮绿黄色，果心中大，心室数4或5，果肉白色，肉质中粗，松脆，汁液中多，味甜，含可溶性固形物14.23%，可滴定酸0.27%；品质中上等。果实较耐贮藏。

38. Xiali

Origin and Distribution Xiali (2n=34), originated in Shanxi Province, is grown mainly in Yuanping, Wutai, and Yuci, etc.

Main Characters Tree: vigorous, productive; late precocity, long-aged. Leaf: 11.2cm × 6.9cm in size, ovate. Initial leaf: red with green. Flower: white bud, 6 to 8 flowers per cluster in average of 6.8; stamen number: 19 to 21 averaging 20.0; corolla diameter: 4.2cm. Fruit: matures in late September in Xingcheng, Liaoning Province, 146g per fruit, 6.3cm long, 6.4cm wide, obovate, greenish-yellow skin, medium core, 4 or 5 locules, flesh white, mid-coarse, crisp tender, mid-juicy, sweet; TSS 14.23%, TA 0.27%; quality above medium; storage life relatively long.

39. 香椿梨

来源及分布 原产陕西省大荔县。

主要性状 树势强，产量中等，始果年龄中或晚。叶片长9.4cm，宽6.0cm，卵圆形，初展叶红色，带有绿色，有茸毛。花蕾白色，边缘浅粉红色，每花序6～8朵花，平均6.7朵；雄蕊19～26枚，平均20.6枚；花冠直径3.9cm。在辽宁兴城，果实9月下旬成熟，单果重147g，纵径6.3cm，横径6.4cm，圆形或扁圆柱形；果皮黄绿色；果心中大，3～5心室；果肉白色，肉质细，稍紧密，脆，汁液中多，味酸甜；含可溶性固形物10.73%，可滴定酸0.29%；品质中上等。果实耐贮藏。

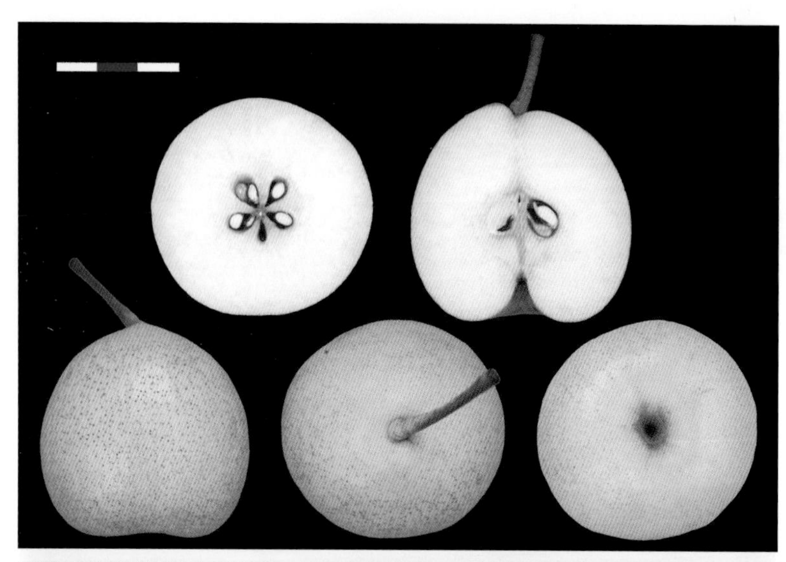

39. Xiangchunli

Origin and Distribution Xiangchunli, was originated in Dali County, Shaanxi Province.

Main Characters Tree: vigorous, medium productive, medium or late precocity. Leaf: 9.4cm × 6.0cm in size, ovate. Initial leaf: red with green, pubescent. Flower: white bud tinged with light pink on edge, 6 to 8 flowers per cluster in average of 6.7; stamen number: 19 to 26, averaging 20.6; corolla diameter: 3.9cm. Fruit: matures in late September in Xingcheng, Liaoning Province, 147g per fruit, 6.3cm long, 6.4cm wide, short cylindrical or round, yellowish-green skin, medium core, 3 to 5 locules, flesh white, fine, crisp, somewhat dense, mid-juicy, sour-sweet; TSS 10.73%, TA 0.29%; quality above medium; storage life long.

40. 西降坞

来源及分布 2n=34，原产安徽省歙县，在江西婺源有栽培。

主要性状 树势强，丰产，始果年龄中等。叶片长 10.6cm，宽 7.7cm，阔卵圆形，初展叶绿色，带有红色。花蕾白色，边缘浅粉红色，每花序 5 ~ 7 朵花，平均 6.0 朵；雄蕊 18 ~ 20 枚，平均 19.0 枚；花冠直径 3.3cm。在辽宁兴城，果实 9 月中旬成熟，单果重 158g，纵径 6.8cm，横径 6.9cm，阔倒卵圆形；果皮绿色或黄绿色；果心中大，5 心室；果肉白色，肉质细，松脆，汁液多，味甜酸；含可溶性固形物 9.83%；品质中上等。

40. Xijiangwu

Origin and Distribution Xijiangwu (2n=34), originated in Shexian, Anhui Province, is grown mainly in Wuyuan County, Jiangxi Province.

Main Characters Tree: vigorous, productive, medium precocity. Leaf: 10.6cm × 7.7cm in size, broadly ovate. Initial leaf: green with red. Flower: white bud tinged with light pink on edge, 5 to 7 flowers per cluster in average of 6.0; stamen number: 18 to 20, averaging 19.0; corolla diameter: 3.3cm. Fruit: matures in mid-September in Xingcheng, Liaoning Province, 158g per fruit, 6.8cm long, 6.9cm wide, broadly obovate, green or yellowish-green skin, medium core, 5 locules, flesh white, fine, crisp tender, juicy, sweet-sour; TSS 9.83%; quality above medium.

41. 兴隆麻梨

来源及分布 2n=34，原产河北省，在天津蓟县、河北兴隆等地有栽培。

主要性状 树势强，丰产，始果年龄中等。叶片长 10.8cm，宽 8.3cm，卵圆形，初展叶绿色，微显红色。花蕾白色，每花序 6 ～ 7 朵花，平均 6.5 朵；雄蕊 19 ～ 22 枚，平均 20.2 枚；花冠直径 4.5cm。在辽宁兴城，果实 9 月上、中旬成熟，单果重 153g，纵径 6.5cm，横径 6.7cm，近圆形或倒卵圆形；果皮绿黄色，梗洼有片锈；果心中大，5 心室；果肉白色，肉质较细，疏松，汁液多，味酸甜或淡甜；含可溶性固形物 11.58%，可滴定酸 0.27%；品质中上等或上等。

41. Xinglong Mali

Origin and Distribution Xinglong Mali (2n=34), originated in Hebei Province, is grown mainly in Jixian of Tianjin, Xinglong of Hebei Province, etc.

Main Characters Tree: vigorous, productive, medium precocity. Leaf: 10.8cm × 8.3cm in size, ovate. Initial leaf: green with less red. Flower: white bud, 6 to 7 flowers per cluster in average of 6.5; stamen number: 19 to 22 averaging 20.2; corolla diameter: 4.5cm. Fruit: matures in early or mid-September in Xingcheng, Liaoning Province, 153g per fruit, 6.5cm long, 6.7cm wide, sub-round or obovate, greenish-yellow skin covered with russet on stalk cavity, medium core, 5 locules, flesh white, relatively fine, tender, juicy, sour-sweet or light sweet; TSS 11.58%, TA 0.27%; good or above medium in quality.

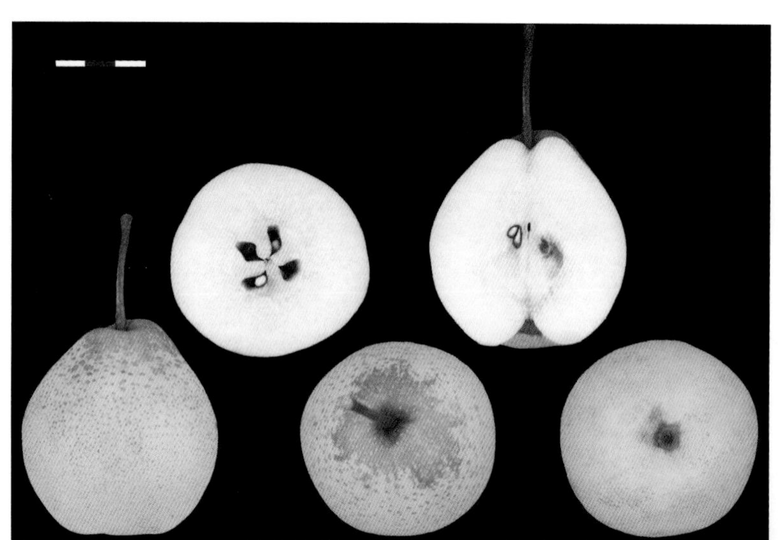

42. 雪花梨

来源及分布 2n=34，原产河北省，以赵县生产的雪花梨最为有名。山西代县、忻县、太原、榆次和陕西渭北各县均有栽培。

主要性状 树势中庸，始果年龄较早，丰产。叶片长 11.4cm，宽 7.7cm，卵圆形，初展叶红色，微显绿色。花蕾白色，边缘淡粉红色，每花序 5 ~ 7 朵花，平均 6.0 朵；雄蕊 19 ~ 24 枚，平均 21.3 枚；花冠直径 4.3cm。在辽宁兴城，果实 9 月下旬成熟，单果重 391g，纵径 9.4cm，横径 8.8cm，长卵圆形或长椭圆形；果皮绿黄色，贮后变黄色；果心小或中大，心室数 5，果肉白色，肉质细，松脆，汁液多，味淡甜，含可溶性固形物 11.60%，可滴定酸 0.11%；品质中上或上等。果实较耐贮藏。

42. Xuehuali

Origin and Distribution Xuehuali (2n=34), was originated in Hebei Province. Fruits produced in Zhaoxian are the most well-known. It is also grown in Dai County, Xin County, Taiyuan, and Yuci of Shanxi Province, as well as counties in Weibei of Shaanxi Province.

Main Characters Tree: moderately vigorous, productive, relatively early precocity. Leaf: 11.4cm × 7.7cm in size, ovate. Initial leaf: red with less green. Flower: white bud tinged with light pink on edge, 5 to 7 flowers per cluster in average of 6.0; stamen number: 19 to 24, averaging 21.3; corolla diameter: 4.3cm. Fruit: matures in late September in Xingcheng, Liaoning Province, 391g per fruit, 9.4cm long, 8.8cm wide, long ovate or oblong, greenish-yellow skin, turning to yellow after storage, small or medium core, 5 locules, flesh white, fine, crisp tender, juicy, light sweet; TSS 11.60%, TA 0.11%; above medium or good in quality; storage life relatively long.

43. 银白梨

来源及分布 原产河北省大名县，为当地古老品种。

主要性状 树势中庸，丰产，始果年龄中或晚。叶片长 10.2cm，宽 7.4cm，卵圆形，初展叶绿色，微有红色。花蕾白色，每花序 6 ~ 8 朵花，平均 7.1 朵；雄蕊 19 ~ 25 枚，平均 22.6 枚；花冠直径 3.6cm。在辽宁兴城，果实 9 月下旬成熟，单果重 121g，纵径 6.3cm，横径 5.0cm，倒卵圆形或长圆形；果皮绿黄色或黄白色；果心中大偏小，5 心室；果肉白色，肉质细，松脆，汁液多，味甜；含可溶性固形物 11.68%，可滴定酸 0.08%；品质中上等。

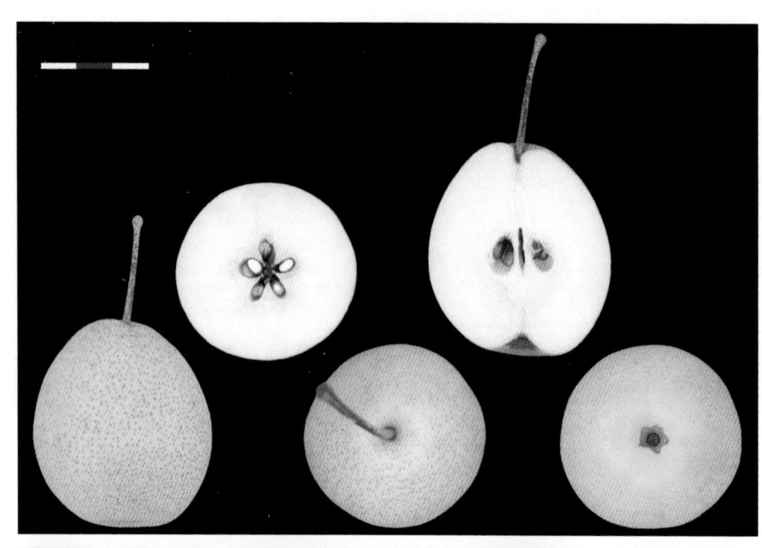

43. Yinbaili

Origin and Distribution Yinbaili, an old pear cultivar, was originated in Daming County, Hebei Province.

Main Characters Tree: moderately vigorous, productive, medium or late precocity. Leaf: 10.2cm × 7.4cm in size, ovate. Initial leaf: green tinged with light red. Flower: white bud, 6 to 8 flowers per cluster in average of 7.1; stamen number: 19 to 25, averaging 22.6; corolla diameter: 3.6cm. Fruit: matures in late September in Xingcheng, Liaoning Province, 121g per fruit, 6.3cm long, 5.0cm wide, obovate or long round, greenish-yellow or whitish-yellow skin, medium core, tending to small, 5 locules, flesh white, fine, crisp tender, juicy, sweet; TSS 11.68%, TA 0.08%; quality above medium.

44. 硬枝青

来源及分布 2n=34，原产江苏省，在睢宁、宿迁等地有栽培。

主要性状 树势中庸，丰产，始果年龄晚。叶片长 11.4cm，宽 7.8cm，卵圆形，初展叶绿色，着红色。花蕾白色，每花序 4 ～ 6 朵花，平均 5.2 朵；雄蕊 21 ～ 28 枚，平均 25.4 枚；花冠直径 3.8cm。在辽宁兴城，果实 9 月下旬成熟，单果重 125g，纵径 6.4cm，横径 6.0cm，阔倒卵圆形或近圆柱形；果皮绿色；果心小，5 心室；果肉白色，肉质细，松脆，汁液多，味淡甜；含可溶性固形物 10.30%，可滴定酸 0.11%；品质中上或上等。

44. Yingzhiqing

Origin and Distribution Yingzhiqing (2n=34), originated in Jiangsu Province, is grown mainly in Suining, Suqian, etc.

Main Characters Tree: moderately vigorous, productive, late precocity. Leaf: 11.4cm × 7.8cm in size, ovate. Initial leaf: green tinged with red. Flower: white bud, 4 to 6 flowers per cluster in average of 5.2; stamen number: 21 to 28, averaging 25.4; corolla diameter: 3.8cm. Fruit: matures in late September in Xingcheng, Liaoning Province, 125g per fruit, 6.4cm long, 6.0cm wide, broadly obovate or sub-cylindrical, green skin, small core, 5 locules, flesh white, fine, crisp tender, juicy, light sweet; TSS 10.30%, TA 0.11%; good or above medium in quality.

45. 鸭老梨

来源及分布　2n=34，原产河北省，在兴隆县一带有少量栽培。

主要性状　树势中庸，较丰产，有隔年结果现象，始果年龄中或晚。叶片长 12.7cm，宽 8.7cm，卵圆形，初展叶绿色，着淡红色。花蕾白色，每花序 5 ～ 9 朵花，平均 7.2 朵；雄蕊 18 ～ 22 枚，平均 19.7 枚；花冠直径 4.3cm。在辽宁兴城，果实 9 月中、下旬成熟，单果重 196g，纵径 7.2cm，横径 7.0cm，长圆形或阔倒卵圆形；果皮绿黄色，梗洼有条锈；果心中大，5 心室；果肉白色，肉质较细，松脆，汁液多，味甜；含可溶性固形物 13.73%，可滴定酸 0.14%；品质上或中上等。果实较耐贮藏。

45. Yalaoli

Origin and Distribution Yalaoli (2n=34), originated in Hebei Province, is grown limitedly in Xinglong, etc.

Main Characters Tree: moderately vigorous, relatively productive, alternate bearing, late or medium precocity. Leaf: 12.7cm × 8.7cm in size, ovate. Initial leaf: green tinged with light red. Flower: white bud, 5 to 9 flowers per cluster in average of 7.2; stamen number: 18 to 22, averaging 19.7; corolla diameter: 4.3cm. Fruit: matures in mid or late September in Xingcheng, Liaoning Province, 196g per fruit, 7.2cm long, 7.0cm wide, long round or obovate, greenish-yellow skin covered with striped russet on stalk cavity, medium core, 5 locules, flesh white, relative fine, crisp tender, juicy, sweet; TSS 13.73%, TA 0.14%; good or above medium in quality; storage life relatively long.

46. 鸭梨

来源及分布 2n=34，原产河北省。河北、山东、山西、陕西、河南栽培最多，辽宁、甘肃、新疆均有栽培。

主要性状 树势中庸或较弱，枝条稀疏屈曲，较丰产，始果年龄早或中等。叶片长13.4cm，宽8.1cm，卵圆形，初展叶红色，微显绿色，有茸毛。花蕾白色，小花蕾边缘淡粉红色，每花序6～9朵花，平均7.9朵；雄蕊20～24枚，平均21.1枚；花冠直径4.2cm。在辽宁兴城，果实9月下旬成熟，单果重159g，纵径7.1cm，横径6.5cm，倒卵圆形，果梗一侧常有突起，并具有锈块；采收时果皮绿黄色，贮后变黄色；果心小，5心室，果肉白色，肉质细，松脆，汁液极多，味淡甜，含可溶性固形物11.98%，可滴定酸0.18%；品质上等。贮藏果实易黑心。

46. Yali

Origin and Distribution Yali (2n=34), originated in Hebei Province, is widely grown in Hebei, Shandong, Shanxi, Shaanxi, and Henan, also cultivated in Liaoning, Gansu, and Xinjiang, etc.

Main Characters Tree: moderately vigorous or weak; sparse and zigzag branch, productive, early or medium precocity. Leaf: 13.4cm × 8.1cm in size, ovate. Initial leaf: red with less green, pubescent. Flower: white bud, small one with light pink on edge, 6 to 9 flowers per cluster in average of 7.9; stamen number: 20 to 24, averaging 21.1; corolla diameter: 4.2cm. Fruit: matures in late September in Xingcheng, Liaoning Province, 159g per fruit, 7.1cm long, 6.5cm wide, obovate, lipped and russet on stalk side, greenish-yellow skin when harvested, turning to yellow after storage, small core, 5 locules, flesh white, fine, crisp tender, extremely juicy, light sweet; TSS 11.98%, TA 0.18%; quality good; core easily browning during storage.

47. 油梨

来源及分布 2n=34，又名香水梨，原产山西省原平县，在原平、五台、榆次、寿阳、平定等县有栽培。

主要性状 树势强，丰产，始果年龄中或晚。叶片长 11.6cm，宽 7.7cm，卵圆形，初展叶绿色，着红色。花蕾白色，小花蕾边缘粉红色，每花序 5 ~ 6 朵花，平均 5.3 朵；雄蕊 22 ~ 34 枚，平均 26.6 枚；花冠直径 4.2cm。在辽宁兴城，果实 9 月下旬成熟，单果重 102g，纵径 5.1cm，横径 5.7cm，扁圆形；果皮绿黄色；果心小，5 心室；果肉白色，肉质细，脆，汁液多，味酸甜；含可溶性固形物 12.54%，可滴定酸 0.20%；品质中上等。果实耐贮藏。

47. Youli

Origin and Distribution Youli (2n=34), also known as Xiangshuili, originated in Yuanping County, Shanxi Province, is grown mainly in Yuanping, Wutai, Yuci, Shouyang, Pingding, etc.

Main Characters Tree: vigorous, productive, medium or late precocity. Leaf: 11.6cm × 7.7cm in size, ovate. Initial leaf: green tinged with red. Flower: white bud, small one tinged with pink on edge, 5 to 6 flowers per cluster in average of 5.3; stamen number: 22 to 34, averaging 26.6; corolla diameter: 4.2cm. Fruit: matures in late September in Xingcheng, Liaoning Province, 102g per fruit, 5.1cm long, 5.7cm wide, oblate, greenish-yellow skin, small core, 5 locules, flesh white, fine, crisp, juicy, sour-sweet; TSS 12.54%, TA 0.20%; quality above medium; storage life long.

48. 早梨

来源及分布 2n=34，产于江西上饶。

主要性状 树势强，始果年龄晚，产量中等。叶片长12.2cm，宽9.1cm，卵圆形，初展叶绿色，着红色。花蕾白色，边缘淡粉红色，每花序6～8朵花，平均6.7朵；雄蕊19～26枚，平均22.0枚；花冠直径3.6cm。在辽宁兴城，果实9月下旬成熟，单果重165g，纵径7.1cm，横径7.5cm，倒卵圆形；果皮绿黄色，有红晕；果心中大，5心室；果肉白色，肉质细，松脆，汁液多，味酸甜，含可溶性固形物11.53%，可滴定酸0.15%；品质中上或上等。

48. Zaoli

Origin and Distribution Zaoli (2n=34), was originated in Shangrao, Jiangxi Province.

Main Characters Tree: vigorous, medium productive, late precocity. Leaf: 12.2cm × 9.1cm in size, ovate. Initial leaf: green tinged with red. Flower: white bud tinged with light pink on edge, 6 to 8 flowers per cluster in average of 6.7; stamen number: 19 to 26, averaging 22.0; corolla diameter: 3.6cm. Fruit: matures in late September in Xingcheng, Liaoning Province, 165g per fruit, 7.1cm long, 7.5cm wide, obovate, greenish-yellow skin tinged with light red on the side exposed to the sun, medium core, 5 locules, flesh white, fine, crisp tender, juicy, sour-sweet; TSS 11.53%, TA 0.15%; above medium or good in quality.

（二）砂梨品种 Sand Pear Varieties

1. 宝珠梨

来源及分布 2n=34，原产云南，在云南呈贡、晋宁等地有栽培。

主要性状 树势强，丰产，始果年龄中或晚。叶片长 9.4cm，宽 6.7cm，卵圆形，初展叶红色，微显绿色，有明显的茸毛。花蕾白色，小花蕾边缘淡粉红色，每花序 5 ～ 7 朵花，平均 5.7 朵；雄蕊 24 ～ 35 枚，平均 29.1 枚；花冠直径 4.8cm。在辽宁兴城，果实 9 月下旬成熟，单果重 198g，纵径 6.9cm，横径 7.2cm，近圆形；果皮黄绿色，厚；果心中大，5 个心室；果肉白色，肉质中粗，松脆，汁液多，味甜；含可溶性固形物 13.63%，可滴定酸 0.23% ；品质上等或中上等。

1. Baozhuli

Origin and Distribution Baozhuli (2n=34), originated in Yunnan Province, is grown mainly in Chenggong, Jinning, etc.

Main Characters Tree: vigorous, productive, medium or late precocity. Leaf: 9.4cm × 6.7cm in size, ovate.Initial leaf: red with less green, pubescent. Flower: white bud, small one tinged with light pink on edge, 5 to 7 flowers per cluster in average of 5.7; stamen number: 24 to 35, averaging 29.1; corolla diameter: 4.8cm. Fruit: matures in late September in Xingcheng, Liaoning Province, 198g per fruit, 6.9cm long, 7.2cm wide, sub-round, yellowish-green skin, thick, medium core, 5 locules, flesh white, mid-coarse, crisp tender, juicy, sweet; TSS 13.63%, TA 0.23%; good or above medium in quality.

2. 大叶雪

来源及分布 2n=3x=51，原产江西婺源。

主要性状 树势中庸，始果年龄中或晚。叶片长 11.3cm，宽 7.6cm，卵圆形，初展叶红色，带有绿色。花蕾白色，边缘浅粉红色，每花序 7 ～ 8 朵花，平均 7.4 朵；雄蕊 20 ～ 23 枚，平均 21.6 枚；花冠直径 5.1cm。在辽宁兴城，果实 9 月上、中旬成熟，单果重 152g，纵径 7.1cm，横径 6.6cm，倒阔圆锥形；果皮绿黄色；果心小，5 个心室；果肉白色，肉质较细，疏松，汁液多，味酸甜；含可溶性固形物 10.43%；品质中上等。

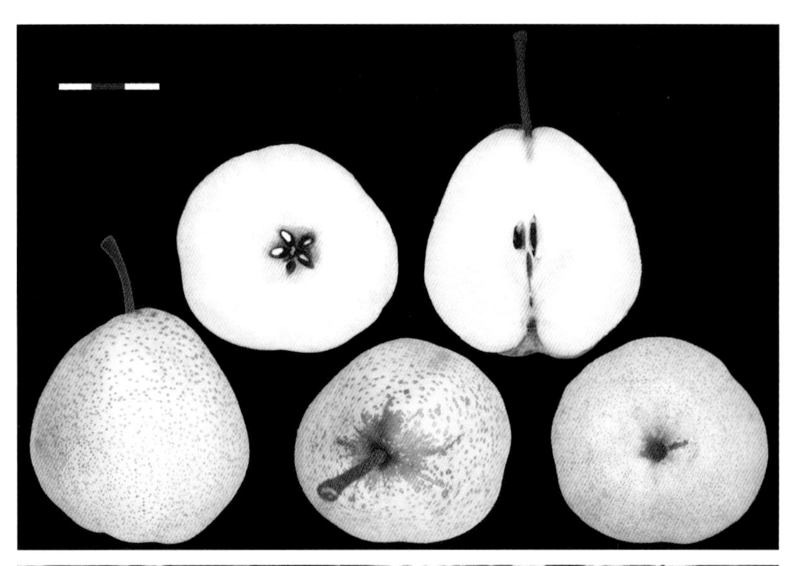

2. Dayexue

Origin and Distribution Dayexue (2n=3x=51), was originated in Wuyuan, Jiangxi Province.

Main Characters Tree: vigorous, medium or late precocity. Leaf: 11.3cm × 7.6cm in size, ovate. Initial leaf: red with green. Flower: white bud tinged with light pink on edge, 7 to 8 flowers per cluster in average of 7.4; stamen number: 20 to 23, averaging 21.6; corolla diameter: 5.1cm. Fruit: matures in early or mid-September in Xingcheng, Liaoning Province, 152g per fruit, 7.1cm long, 6.6cm wide, upside-down broadly conical, greenish-yellow skin, small core, 5 locules, flesh white, relatively fine, tender, juicy, sour-sweet; TSS 10.43%; quality above medium.

3. 苍梧大沙梨

来源及分布　2n=34，原产广西梧州。

主要性状　树势强，丰产，始果年龄中等。叶片长 9.2cm，宽 6.9cm，卵圆形，初展叶绿色，带有红色。花蕾白色，每花序 5 ～ 8 朵花，平均 6.7 朵；雄蕊 18 ～ 26 枚，平均 23.3 枚。在湖北武汉，果实 9 月下旬成熟，单果重 375g，纵径 8.2cm，横径 9.5cm，阔卵圆形或阔圆锥形；果皮绿褐色或黄褐色，厚；果心小，5 心室；果肉白色，肉质细，松脆，汁液多，味甜或淡甜；含可溶性固形物 12.0%，可滴定酸 0.13%；品质中上等或上等。

3. Cangwu Dashali

Origin and Distribution Cangwu Dashali (2n=34), was originated in Wuzhou, Guangxi Province.

Main Characters Tree: vigorous, productive, medium precocity. Leaf: 9.2cm × 6.9cm in size, ovate. Initial leaf: green tinged with red. Flower: white bud, 5 to 8 flowers per cluster in average of 6.7; stamen number: 18 to 26, averaging 23.3. Fruit: matures in late September in Wuhan, Hubei Province, 375g per fruit, 8.2cm long, 9.5cm wide, broadly ovate or broadly conical, greenish-brown or yellowish-brown skin, thick, small core, 5 locules, flesh white, fine, crisp tender, juicy, sweet or light sweet; TSS 12.0%, TA 0.13%; good or above medium in quality.

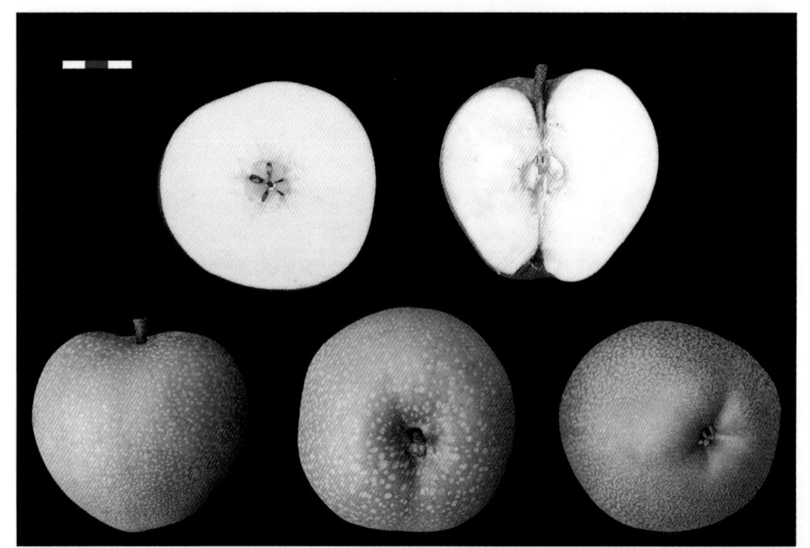

4. 苍溪雪梨

来源及分布 2n=34，又名施家梨或苍溪梨，原产于四川省苍溪县，为砂梨系统著名地方品种之一。

主要性状 树势中庸，丰产，始果年龄中等。叶片长11.2cm，宽6.9cm，狭椭圆形，初展叶绿色，着浅红色。花蕾白色，小花蕾边缘浅粉红色，每花序6～7朵花，平均6.8朵；雄蕊19～20枚，平均19.9枚；花冠直径4.5cm。在辽宁兴城，果实9月下旬成熟，单果重321g，纵径9.3cm，横径8.1cm，长倒卵圆形或瓢形；果皮褐色；果心小，5心室；果肉淡绿白色，肉质中粗，疏松，汁液多，味甜或淡甜；含可溶性固形物11.43%，可滴定酸0.11%；品种上等或中上等。

4. Cangxi Xueli

Origin and Distribution Cangxi Xueli (2n=34), also known as Shijiali, or Cangxili, one of famous Sand Pear cultivars, was originated in Cangxi County, Sichuan Province.

Main Characters Tree: moderately vigorous, productive, medium precocity. Leaf: 11.2cm × 6.9cm in size, narrowly ovate. Initial leaf: green tinged with light red. Flower: white bud, small one tinged with light pink on edge, 6 to 7 flowers per cluster in average of 6.8; stamen number: 19 to 20, averaging 19.9; corolla diameter: 4.5cm. Fruit: matures in late September in Xingcheng, Liaoning Province, 321g per fruit, 9.3cm long, 8.1cm wide, long obovate or pyriform, brown skin, small core, 5 locules, flesh pale greenish-white, mid-coarse, tender, juicy, sweet or light sweet; TSS 11.43%, TA 0.11%; good or above medium in quality.

5. 洞冠梨

来源及分布 2n=34，原产广东省阳山县洞冠村。

主要性状 树势强，丰产，始果年龄中等。叶片长 8.9cm，宽 5.7cm，卵圆形，初展叶绿色，着红色。花蕾白色，每花序 3 ~ 8 朵花，平均 6.1 朵；雄蕊 18 ~ 31 枚，平均 22.9 枚。在湖北武汉，果实 9 月上旬成熟，单果重 581g，纵径 9.3cm，横径 10.7cm，阔圆锥形或阔卵圆形；果皮褐色；果心极小，5 心室；果肉白色，肉质细，脆，汁液多，味甜；含可溶性固形物 8.7%，可滴定酸 0.19%；品质中上等。

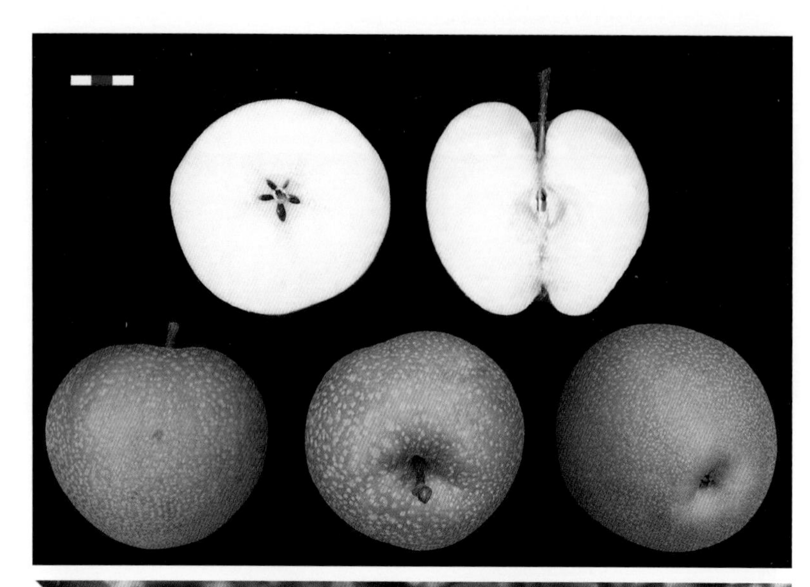

5. Dongguanli

Origin and Distribution Dongguanli (2n=34), was originated in Dongguan Village, Yangshan County, Guangdong Province.

Main Characters Tree: vigorous, productive, medium precocity. Leaf: 8.9cm × 5.7cm in size, ovate. Initial leaf: green tinged with red. Flower: white bud, 3 to 8 flowers per cluster in average of 6.1; stamen number: 18 to 31, averaging 22.9. Fruit: matures in early September in Wuhan, Hubei Province, 581g per fruit, 9.3cm long, 10.7cm wide, broadly conical or broadly ovate, brown skin, extremely small core, 5 locules, flesh white, fine, crisp, juicy, sweet; TSS 8.7%, TA 0.19%; quality above medium.

6. 灌阳雪梨

来源及分布 2n=34，原产中国，在广西灌阳有栽培，据传由天津或四川引入。

主要性状 树势中庸，丰产，始果年龄中等。叶片长 11.8cm，宽 7.6cm，阔卵圆形，初展叶绿色。花蕾白色，边缘粉红色，每花序 7 ~ 9 朵花，平均 7.8 朵；雄蕊 19 ~ 20 枚，平均 19.8 枚；花冠直径 3.9cm。在辽宁兴城，果实 9 月中旬成熟，单果重 168g，纵径 7.1cm，横径 6.6cm，倒卵圆形或长圆形；果皮黄褐色或绿褐色；果心中大偏大，5 心室，果肉白色，肉质细，松脆，汁液多，味酸甜；含可溶性固形物 14.77%，可滴定酸 0.21%；品质上等或中上等。

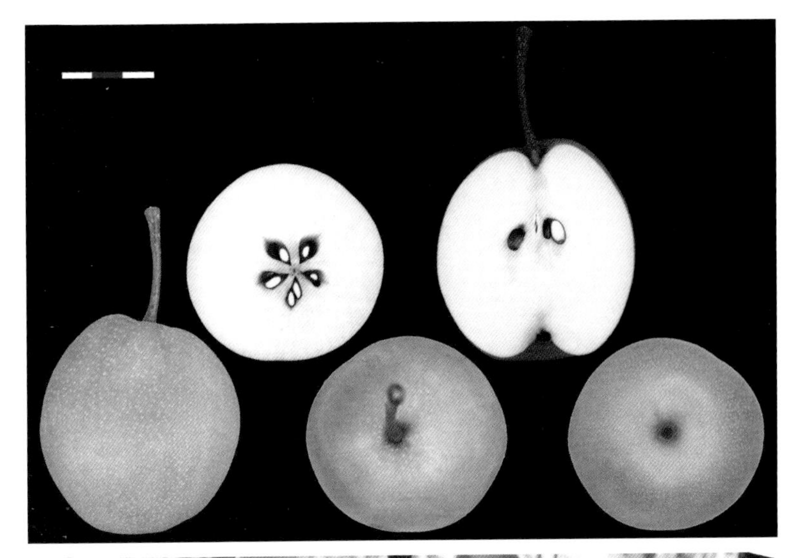

6. Guanyang Xueli

Origin and Distribution Guanyang Xueli (2n=34), originated in China, is grown mainly in Guanyang County, Guangxi Province. It was said to be introduced to Guanyang from Tianjin or Sichuan Province.

Main Characters Tree: moderately vigorous, productive, medium precocity. Leaf: 11.8cm × 7.6cm in size, broadly ovate. Initial leaf: green. Flower: white bud tinged with pink on edge, 7 to 9 flowers per cluster in average of 7.8; stamen number: 19 to 20, averaging 19.8; corolla diameter: 3.9cm. Fruit: matures in mid-September in Xingcheng, Liaoning Province, 168g per fruit, 7.1cm long, 6.6cm wide, obovate or long round, yellowish-brown or greenish-brown skin, medium core, tending to large, 5 locules, flesh white, fine, crisp tender, juicy, sour-sweet; TSS 14.77%, TA 0.21%; good or above medium in quality.

7. 横县蜜梨

来源及分布　2n=34，原产广西横县。

主要性状　树势弱，丰产，始果年龄早。叶片长11.5cm，宽5.8cm，卵圆形，初展叶绿色，着红色。花蕾白色，每花序4～11朵花，平均7.6朵；雄蕊18～25枚，平均21.5枚。在湖北武汉，果实9月中旬成熟，单果重130g，纵径6.3cm，横径6.7cm，扁圆形；果皮黄褐色；果心中大，5心室；果肉淡黄白色，肉质粗，紧密，汁液中多，味甜；含可溶性固形物12.1%，可滴定酸0.20%；品质中上等。

7. Hengxian Mili

Origin and Distribution Hengxian Mili (2n=34), was originated in Hengxian, Guangxi Province.

Main Characters Tree: weak, productive, early precocity. Leaf: 11.5cm × 5.8cm in size, ovate. Initial leaf: green tinged with red. Flower: white bud, 4 to 11 flowers per cluster in average of 7.6; stamen number: 18 to 25, averaging 21.5. Fruit: matures in mid September in Wuhan, Hubei Province, 130g per fruit, 6.3cm long, 6.7cm wide, oblate, russet skin, medium core, 5 locules, flesh pale yellowish-white, coarse, dense, mid-juicy, sweet; TSS 12.1%, TA 0.20%; quality above medium.

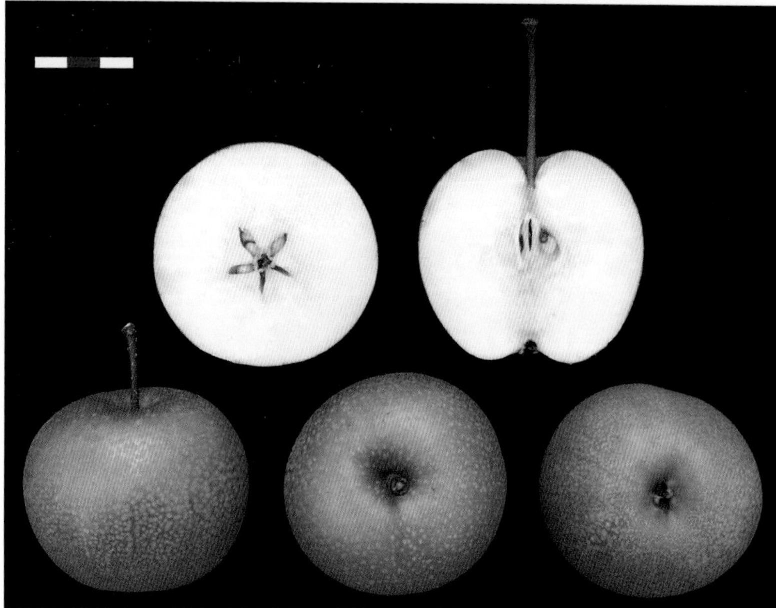

8. 红苕棒梨

来源及分布　2n=34，原产四川简阳。

主要性状　树势强，丰产。叶片长13.0cm，宽8.3cm，卵圆形，初展叶红色。花蕾白色，小花蕾边缘淡粉红色，每花序6～8朵花，平均6.6朵；雄蕊18～21枚，平均19.7枚；花冠直径4.1cm。在辽宁兴城，果实9月中、下旬成熟，单果重268g，纵径8.0cm，横径7.7cm，椭圆形；果皮绿黄色，从梗洼向下有大面积褐色果锈；果心小或中大，5心室；果肉淡黄白色，肉质中粗，脆，汁液多，味甜酸；含可溶性固形物14.33%，可滴定酸0.74%；品质中上等。

8. Hongshaobangli

Origin and Distribution Hongshaobangli (2n=34), was originated in Jianyang, Sichuan Province.

Main Characters Tree: vigorous, productive. Leaf: 13.0cm × 8.3cm in size, ovate. Initial leaf: red. Flower: white bud, small one tinged with light pink on edge, 6 to 8 flowers per cluster in average of 6.6; stamen number: 18 to 21, averaging 19.7; corolla diameter: 4.1cm. Fruit: matures in mid or late September in Xingcheng, Liaoning Province, 268g per fruit, 8.0cm long, 7.7cm wide, elliptical, greenish-yellow skin covered with russet from stalk cavity down to eye basin, small or medium core, 5 locules, flesh pale yellowish-white, mid-coarse, crisp, juicy, sweet-sour; TSS 14.33%, TA 0.74%; quality above medium.

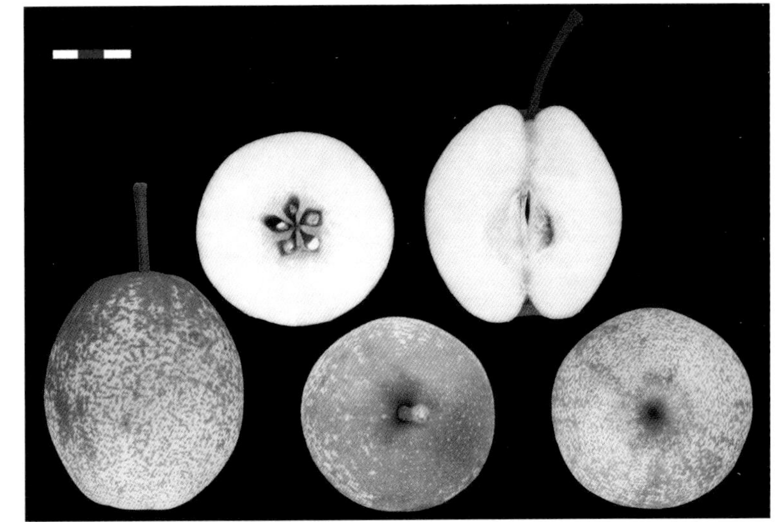

9. 惠水金盖梨

来源及分布 原产贵州省惠水县。

主要性状 树势强，较丰产，始果年龄晚。叶片长 15.2cm，宽 9.8cm，阔卵圆形，初展叶红色。花蕾白色，每花序 6 ~ 9 朵花，平均 6.9 朵；雄蕊 20 ~ 24 枚，平均 21.8 枚；花冠直径 5.0cm。在辽宁兴城，果实 9 月下旬或 10 月上旬成熟，单果重 160g，纵径 6.5cm，横径 6.5cm，近圆形或卵圆形；果皮黄绿色，梗洼有锈，果皮厚；果心中大，5 心室；果肉白色，肉质中粗或粗，松脆，汁液多，味甜或酸甜；含可溶性固形物 12.50%，可滴定酸 0.17%；品质中等或中上等。

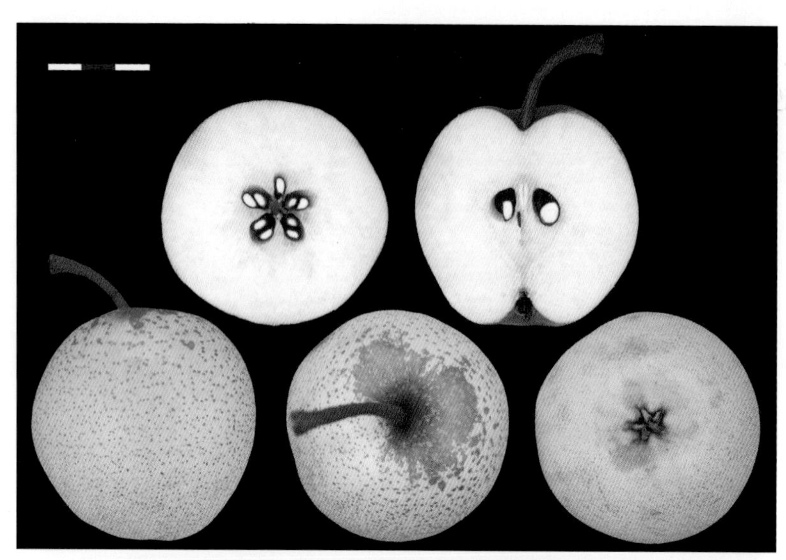

9. Huishui Jingaili

Origin and Distribution Huishui Jingaili, was originated in Huishui, Guizhou Province.

Main Characters Tree: vigorous, relatively productive, late precocity. Leaf: 15.2cm × 9.8cm in size, broadly ovate. Initial leaf: red. Flower: white bud, 6 to 9 flowers per cluster in average of 6.9; stamen number: 20 to 24, averaging 21.8; corolla diameter: 5.0cm. Fruit: matures in late September or early October in Xingcheng, Liaoning Province, 160g per fruit, 6.5cm long, 6.5cm wide, sub-round or ovate, yellowish-green skin with russet in stalk cavity, thick, medium core, 5 locules, flesh white, mid-coarse or coarse, crisp tender, juicy, sweet or sour-sweet; TSS 12.50%, TA 0.17%; quality medium or above.

10. 惠阳红梨

来源及分布 2n=34，原产广东惠阳。

主要性状 树势中庸，丰产。叶片长10.8cm，宽6.6cm，卵圆形，初展叶红色。花蕾白色，边缘粉红色，每花序6～10朵花，平均8.5朵；雄蕊18～25枚，平均21.5枚；花冠直径4.3cm。在辽宁兴城，果实9月下旬成熟，单果重86g，纵径4.8cm，横径5.6cm，扁圆形或扁圆锥形；果皮黄褐色；果心中大，4～6心室；果肉淡黄白色，肉质较细，松脆，汁液中多，味酸甜；含可溶性固形物11.60%；品质中上等。果实耐贮藏。

10. Huiyang Hongli

Origin and Distribution Huiyang Hongli (2n=34), was originated in Huiyang, Guangdong Province.

Main Characters Tree: moderately vigorous, productive. Leaf: 10.8cm × 6.6cm in size, ovate. Initial leaf: red. Flower: white bud tinged with pink on edge, 6 to 10 flowers per cluster in average of 8.5; stamen number: 18 to 25, averaging 21.5; corolla diameter: 4.3cm. Fruit: matures in late September in Xingcheng, Liaoning Province, 86g per fruit, 4.8cm long, 5.6cm wide, oblate or short conical, yellowish-brown skin, medium core, 4 to 6 locules, flesh pale yellowish-white, relatively fine, crisp tender, mid-juicy, sour-sweet; TSS 11.60%; quality above medium; storage life long.

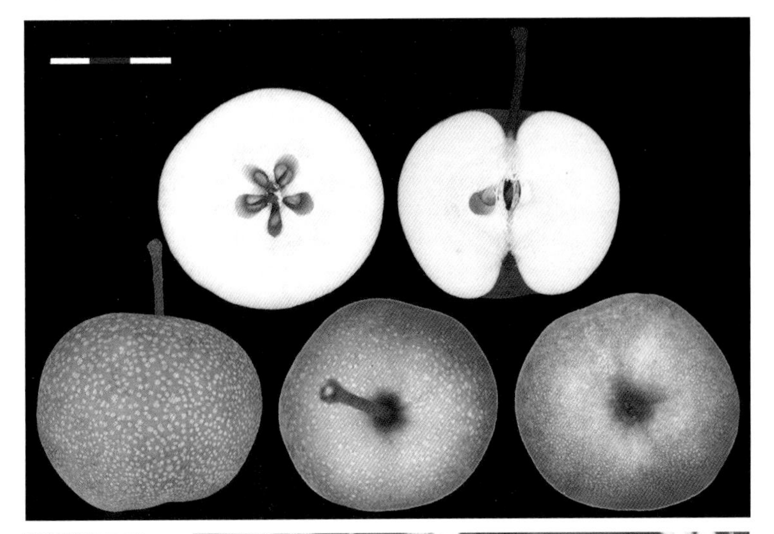

11. 火把梨

来源及分布 2n=34，原产云南，在大理、丽江等地有栽培。

主要性状 树势中庸，丰产性中等，始果年龄晚。叶片长9.2cm，宽4.9cm，狭椭圆形，初展叶绿色，着红色。花蕾白色，边缘浅粉红色，每花序3～8朵花，平均6.0朵；雄蕊26～34枚，平均31.0枚。在湖北武汉，果实9月中、下旬成熟，单果重171g，纵径6.9cm，横径6.8cm，卵圆形或椭圆形；果皮黄绿色，部分果面着鲜红色；果心中大，5心室；果肉黄白色，肉质中粗，松软，汁液多，味甜酸；含可溶性固形物15.0%，可滴定酸0.60%；品质中等。

11. Huobali

Origin and Distribution Huobali (2n=34), originated in Yunnan Province, is grown in Dali, Lijiang, etc.

Main Characters Tree: moderately vigorous, moderately productive, late precocity. Leaf: 9.2cm × 4.9cm in size, narrowly elliptical. Initial leaf: green tinged with red. Flower: white bud tinged with light pink on edge, 3 to 8 flowers per cluster in average of 6.0; stamen number: 26 to 34, averaging 31.0. Fruit: matures in mid or late September in Wuhan, Hubei Province, 171g per fruit, 6.9cm long, 6.8cm wide, ovate or elliptical, yellowish-green skin, some covered with bright red on the side exposed to the sun, medium core, 5 locules, flesh yellowish-white, mid-coarse, soft tender, juicy, sweet-sour; TSS 15.0%, TA 0.60%; quality medium.

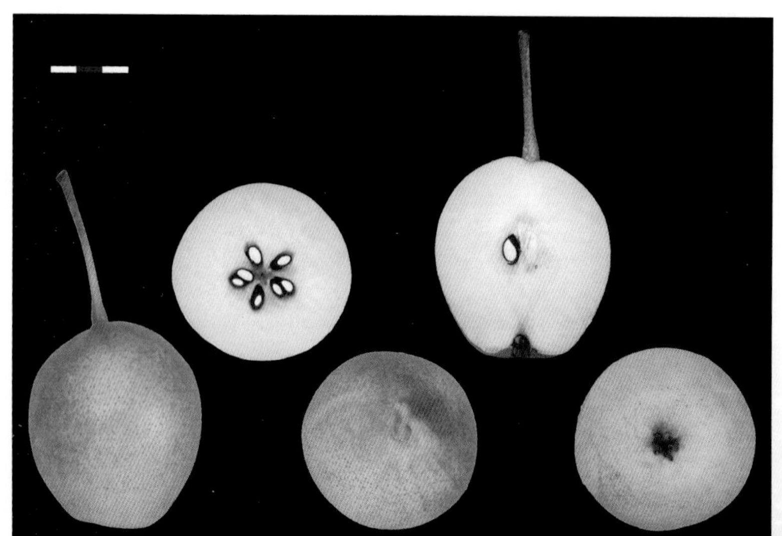

12. 金花早

来源及分布 2n=34，原产安徽歙县。

主要性状 树势弱，丰产，始果年龄中等。叶片长 10.8cm，宽 5.8cm，卵圆形，初展叶红色。花蕾白色，边缘浅粉红色，每花序 4 ～ 7 朵花，平均 5.7 朵；雄蕊 19 ～ 20 枚，平均 19.8 枚。在湖北武汉，果实 9 月上、中旬成熟，单果重 165g，纵径 6.4cm，横径 6.9cm，扁圆形；果皮黄绿色，梗端有片锈；果心小，5 心室；果肉白色，肉质细，松脆，汁液多，味淡甜；含可溶性固形物 9.8%，可滴定酸 0.21%；品质中上等。

12. Jinhuazao

Origin and Distribution Jinhuazao (2n=34), was originated in Shexian, Anhui Province.

Main Characters Tree: weak, productive, medium precocity. Leaf: 10.8cm × 5.8cm in size, ovate. Initial leaf: red. Flower: white bud tinged with light pink on edge, 4 to 7 flowers per cluster in average of 5.7; stamen number: 19 to 20, averaging 19.8. Fruit: matures in early or mid September in Wuhan, Hubei Province, 165g per fruit, 6.4cm long, 6.9cm wide, oblate, yellowish-green skin covered with russet on the stalk end, small core, 5 locules, flesh white, fine, crisp tender, juicy, light sweet; TSS 9.8%, TA 0.21%; quality above medium.

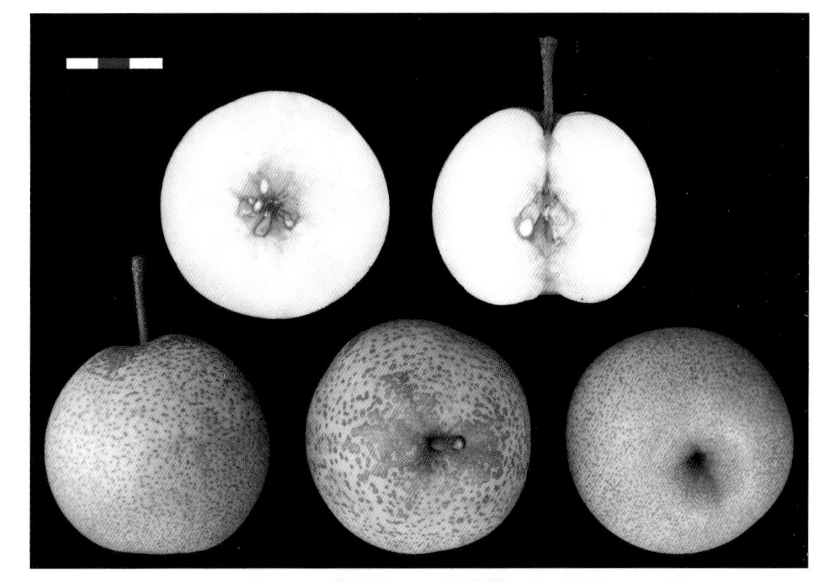

13. 魁星麻壳

来源及分布 2n=34，原产江西上饶。

主要性状 树势弱，丰产，始果年龄晚。叶片长 9.0cm，宽 5.6cm，卵圆形，初展叶绿色，着红色。花蕾白色，边缘浅粉红色，每花序 3 ～ 9 朵花，平均 7.1 朵；雄蕊 17 ～ 23 枚，平均 20.0 枚。在湖北武汉，果实 9 月中旬成熟，单果重 398g，纵径 8.3cm，横径 9.3cm，扁圆形，不规则，果面粗糙；果皮黄绿色；果心中大，5 心室；果肉白色，肉质中粗，松脆，汁液多，味淡甜；含可溶性固形物 7.8%，可滴定酸 0.10%；品质中上等。

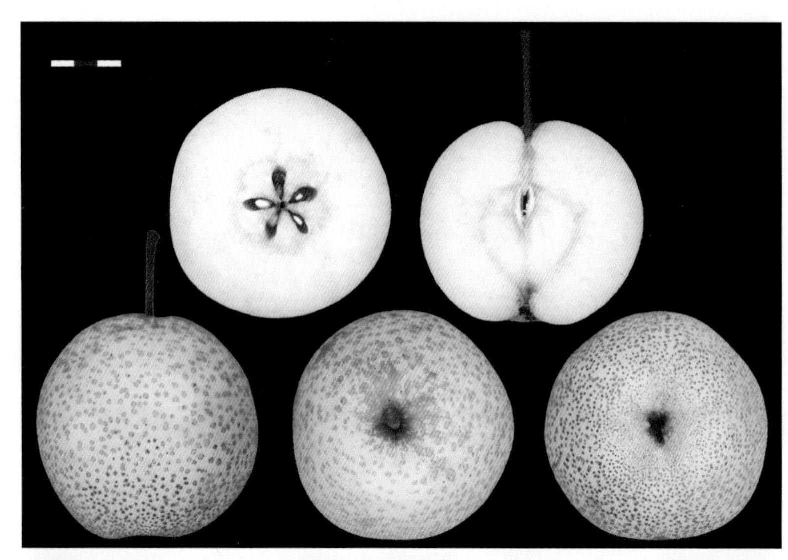

13. Kuixing Make

Origin and Distribution Kuixing Make (2n=34), was originated in Shangrao, Jiangxi Province.

Main Characters Tree: weak, productive, late precocity. Leaf: 9.0cm × 5.6cm in size, ovate. Initial leaf: green tinged with red. Flower: white bud tinged with light pink on edge, 3 to 9 flowers per cluster in average of 7.1; stamen number: 17 to 23, averaging 20.0. Fruit: matures in mid-September in Wuhan, Hubei Province, 398g per fruit, 8.3cm long, 9.3cm wide, oblate, irregular, surface coarse, yellowish-green skin, medium core, 5 locules, flesh white, mid-coarse, crisp tender, juicy, light sweet; TSS 7.8%, TA 0.10%; quality above medium.

14. 利川香水梨

来源及分布 2n=34，原产湖北利川。

主要性状 树势中庸，丰产性中等，始果年龄中等。叶片长 8.7cm，宽 6.3cm，卵圆形，初展叶绿色，着红色。花蕾白色，边缘浅粉红色，每花序 4 ～ 7 朵花，平均 5.7 朵；雄蕊 19 ～ 26 枚，平均 21.3 枚。在湖北武汉，果实 8 月中、下旬成熟，单果重 188g，纵径 7.1cm，横径 6.8cm，阔倒卵圆形；果皮黄绿色；果心中大，5 心室；果肉白色，肉质中粗，松脆，汁液多，味淡甜；含可溶性固形物 10.8%，可滴定酸 0.18%；品质中等。

14. Lichuan Xiangshuili

Origin and Distribution Lichuan Xiangshuili (2n=34), was originated in Lichuan, Hubei Province.

Main Characters Tree: moderately vigorous and productive, medium precocity. Leaf: 8.7cm× 6.3cm in size, ovate. Initial leaf: green tinged with red. Flower: white bud tinged with light pink on edge, 4 to 7 flowers per cluster in average of 5.7; stamen number: 19 to 26, averaging 21.3. Fruit: matures in mid or late August in Wuhan, Hubei Province, 188g per fruit, 7.1cm long, 6.8cm wide, broadly obovate, yellowish-green skin, medium core, 5 locules, flesh white, mid-coarse, crisp tender, juicy, light sweet; TSS 10.8%, TA 0.18%; quality medium.

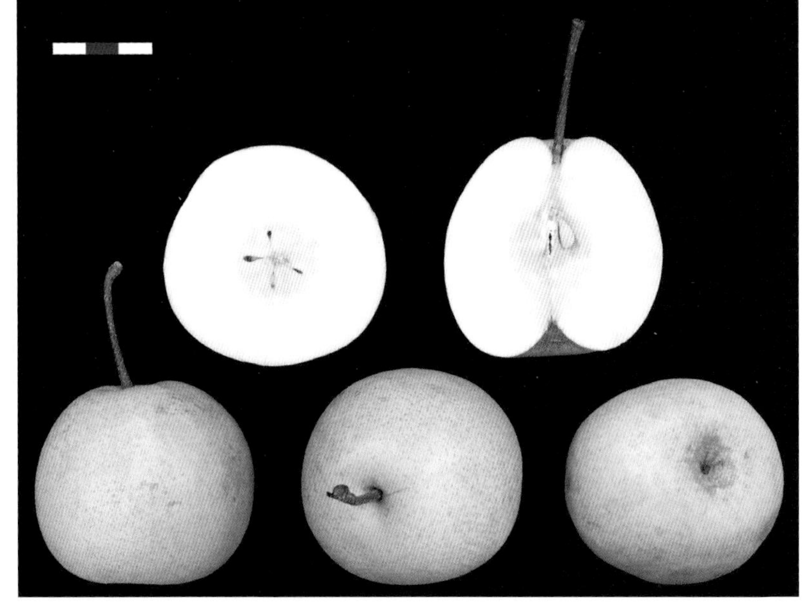

15. 隆回巨梨

来源及分布　2n=34，原产湖南隆回。

主要性状　树势强，丰产，始果年龄中等。叶片长 9.5cm，宽 6.4cm，卵圆形，初展叶红色。花蕾白色，边缘浅粉红色，每花序 4 ~ 7 朵花，平均 5.1 朵；雄蕊 20 ~ 27 枚，平均 23.8 枚。在湖北武汉，果实 10 月中、下旬成熟，单果重 445g，纵径 10.5cm，横径 9.0cm，倒卵圆形；果皮黄绿色，厚；果心极小，5 ~ 6 心室；果肉白色，肉质中粗，疏松，汁液多，味甜；含可溶性固形物 10.6%，可滴定酸 0.18%；品质中上等。

15. Longhui Juli

Origin and Distribution Longhui Juli (2n=34), was originated in Longhui, Hunan Province.

Main Characters Tree: vigorous, productive, medium precocity. Leaf: 9.5cm ×6.4cm in size, ovate. Initial leaf: red. Flower: white bud tinged with light pink on edge, 4 to 7 flowers per cluster in average of 5.1; stamen number: 20 to 27, averaging 23.8. Fruit: matures in mid or late October in Wuhan, Hubei Province, 445g per fruit, 10.5cm long, 9.0cm wide, obovate, yellowish-green skin, thick, extremely small core, 5 or 6 locules, flesh white, mid-coarse, tender, juicy, sweet; TSS 10.6%, TA 0.18%; quality above medium.

16. 满顶雪

来源及分布 2n=34，原产福建浦城。

主要性状 树势强，产量中等，始果年龄早或中等。叶片长9.8cm，宽7.1cm，卵圆形，初展叶绿色，着浅红色。花蕾白色，小花蕾边缘着浅粉红色，每花序5～8朵花，平均6.7朵；雄蕊24～33枚，平均28.6枚；花冠直径4.0cm。在辽宁兴城，果实9月下旬成熟，单果重225g，纵径7.0cm，横径7.6cm，阔倒卵圆形；果皮褐色；果心小，5心室；肉质较细，脆，汁液中多，味酸甜或甜；含可溶性固形物13.40%，可滴定酸0.18%；品质中上等或上等。

16. Mandingxue

Origin and Distribution Mandingxue (2n=34), was originated in Pucheng, Fujian Province.

Main Characters Tree: vigorous, medium productive, early or medium precocity. Leaf: 9.8cm × 7.1cm in size, ovate. Initial leaf: green tinged with light red. Flower: white bud, small one tinged with light pink on edge, 5 to 8 flowers per cluster in average of 6.7; stamen number: 24 to 33, averaging 28.6; corolla diameter: 4.0cm. Fruit: matures in late September in Xingcheng, Liaoning Province, 225g per fruit, 7.0cm long, 7.6cm wide, broadly obovate, brown skin, small core, 5 locules, flesh relatively fine, crisp, mid-juicy, sour-sweet or sweet; TSS 13.40%, TA 0.18%; good or above medium in quality.

17. 糯稻梨

来源及分布　2n=34，原产浙江义乌。

主要性状　树势强，丰产，始果年龄早或中等。叶片长 9.0cm，宽 6.1cm，广卵圆形，初展叶绿色，着红色。花蕾白色，小花蕾边缘着浅粉红色，每花序 5 ~ 7 朵花，平均 6.1 朵；雄蕊 20 ~ 21 枚，平均 20.1 枚；花冠直径 2.8cm。在辽宁兴城，果实 9 月中、下旬成熟，单果重 223g，纵径 7.7cm，横径 7.2cm，长圆形；果皮绿黄色或黄白色，梗洼有锈斑；果心小或中大，4 或 5 心室；果肉白色，肉质中粗，疏松，汁液多，味淡甜或甜；含可溶性固形物 14.60%，可滴定酸 0.14%；品质中上等或中等。

17. Nuodaoli

Origin and Distribution Nuodaoli (2n=34), was originated in Yiwu, Zhejiang Province

Main Characters Tree: vigorous, productive, early or medium precocity. Leaf: 9.0cm × 6.1cm in size, broadly ovate. Initial leaf: green tinged with red. Flower: white bud, small one tinged with light pink on edge, 5 to 7 flowers per cluster in average of 6.1; stamen number: 20 to 21, averaging 20.1; corolla diameter: 2.8cm. Fruit: matures in mid or late September in Xingcheng, Liaoning Province, 223g per fruit, 7.7cm long, 7.2cm wide, long round, greenish-yellow or whitish-yellow skin covered with russet on stalk cavity, small or medium core, 4 or 5 locules, flesh white, mid-coarse, tender, juicy, light sweet or sweet; TSS 14.60%, TA 0.14%; quality medium or above.

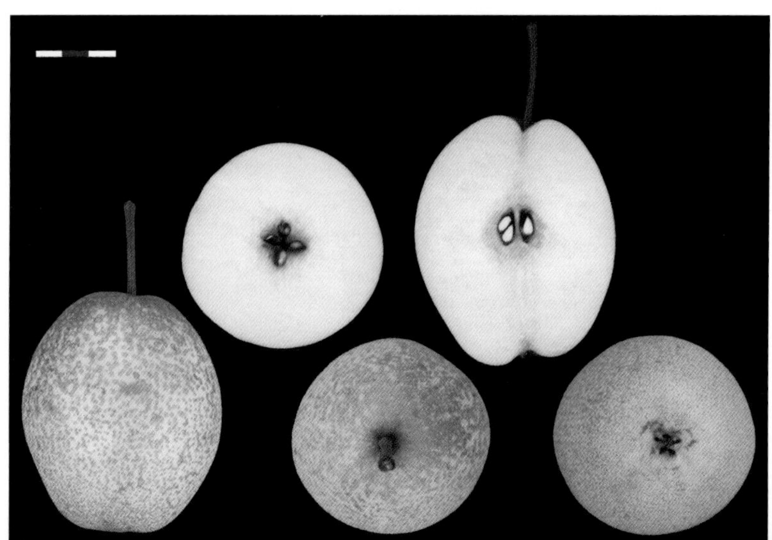

18. 蒲瓜梨

来源及分布 又名大恩梨、早梨，原产浙江乐清。

主要性状 树势强，较丰产。叶片长9.2cm，宽8.2cm，近圆形，初展叶绿色，着浅红色。花蕾白色，每花序6～8朵花，平均7.0朵；雄蕊20～22枚，平均20.2枚；花冠直径3.7cm。在辽宁兴城，果实9月下旬成熟，单果重261g，纵径8.5cm，横径7.7cm，阔倒卵圆形或阔纺锤形；果皮黄绿色或黄色，梗端有片锈；果心极小，5心室；果肉白色，肉质较细，松脆，汁液多，味酸甜；含可溶性固形物11.87%，可滴定酸0.27%；品质中上等。果实较耐贮藏。

18. Puguali

Origin and Distribution Puguali, also known as Daenli, or Zaoli, was originated in Leqing, Zhejiang Province.

Main Characters Tree: vigorous, relatively productive. Leaf: 9.2cm × 8.2cm in size, sub-round. Initial leaf: green tinged with light red. Flower: white bud, 6 to 8 flowers per cluster in average of 7.0; stamen number: 20 to 22, averaging 20.2; corolla diameter: 3.7cm. Fruit: matures in late September in Xingcheng, Liaoning Province, 261g per fruit, 8.5cm long, 7.7cm wide, broadly obovate or spindle-shaped, yellowish-green or yellow skin covered with russet on stalk end, extremely small core, 5 locules, flesh white, relatively fine, crisp tender, juicy, sour-sweet; TSS 11.87%, TA 0.27%; quality above medium. storage life relatively long.

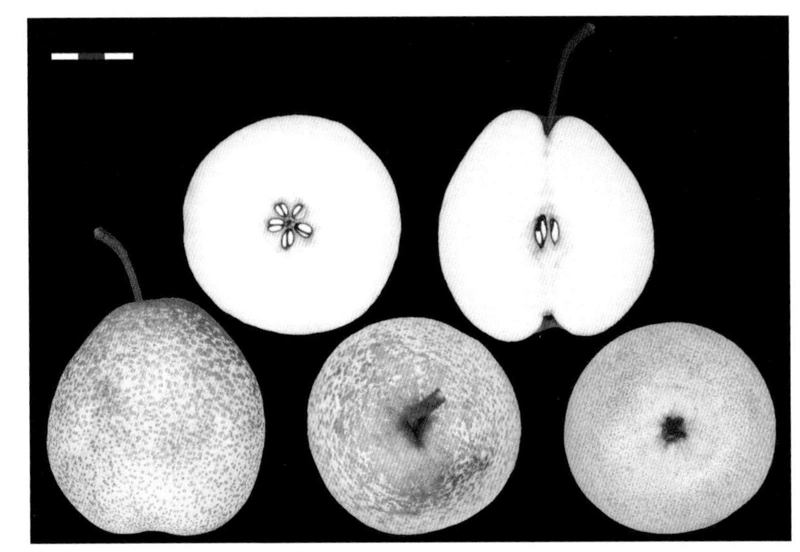

19. 蒲梨宵

来源及分布 2n=34，原产江西南昌。

主要性状 树势中庸，丰产，始果年龄晚。叶片长 11.1cm，宽 11.2cm，圆形或阔卵圆形，初展叶褐红色。花蕾白色，边缘粉红色，每花序 3 ~ 8 朵花，平均 5.9 朵；雄蕊 18 ~ 27 枚，平均 21.0 枚。在湖北武汉，果实 9 月上旬成熟，单果重 240g，纵径 7.5cm，横径 7.5cm，阔倒卵圆形或扁圆形；果皮绿黄色或黄色；果心极小，5 心室；果肉白色，肉质细，松脆，汁液多，味酸甜或淡甜；含可溶性固形物 10.1%，可滴定酸 0.30%；品质中上等。

19. Pulixiao

Origin and Distribution Pulixiao (2n=34), was originated in Nanchang, Jiangxi Province.

Main Characters Tree: moderately vigorous, productive, late precocity. Leaf: 11.1cm ×11.2cm in size, round or broadly ovate. Initial leaf: brownish-red. Flower: white bud tinged with pink on edge, 3 to 8 flowers per cluster in average of 5.9; stamen number: 18 to 27, averaging 21.0. Fruit: matures in early September in Wuhan, Hubei Province, 240g per fruit, 7.5cm long, 7.5cm wide, broadly obovate or oblate, greenish-yellow or yellow skin, extremely small core, 5 locules, flesh white, fine, crisp tender, juicy, sour-sweet or light sweet; TSS 10.1%, TA 0.30%; quality above medium.

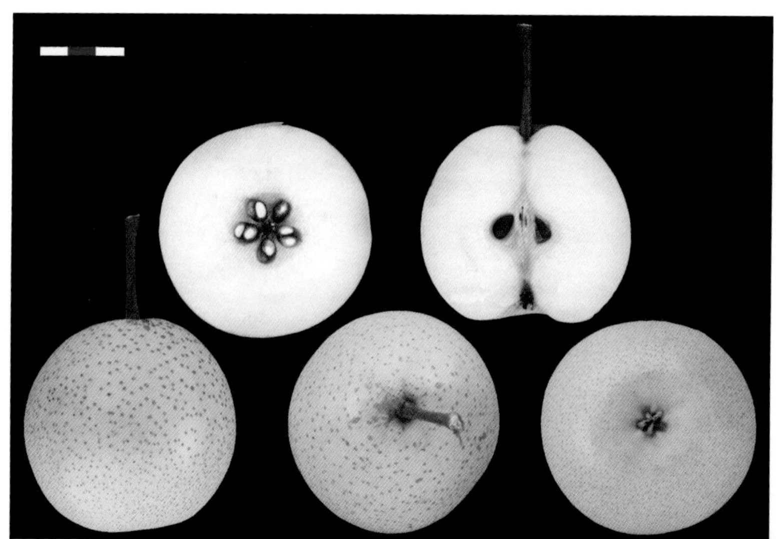

20. 嵊县秋白梨

来源及分布 又名白樟梨，原产浙江嵊县。

主要性状 树势强，丰产，始果年龄中或晚。叶片长 11.1cm，宽 7.5cm，卵圆形，初展叶绿色，着浅红色。花蕾白色，边缘淡粉红色，每花序 5 ~ 7 朵花，平均 6.1 朵；雄蕊 17 ~ 20 枚，平均 19.4 枚；花冠直径 3.8cm。在辽宁兴城，果实 9 月下旬成熟，单果重 118g，纵径 5.9cm，横径 5.7cm，近圆形；果皮较厚，绿黄色或黄白色；果心中大，5 心室；果肉浅黄白色，肉质脆，中粗，汁液多，味酸甜或淡甜；含可溶性固形物 13.3%，可滴定酸 0.11%；品质中上等。果实耐贮藏。

20. Shengxian Qiubaili

Origin and Distribution Shengxian Qiubaili, also known as Baizhangli, was originated in Shengxian, Zhejiang Province.

Main Characters Tree: vigorous, productive, medium or late precocity. Leaf: 11.1cm × 7.5cm in size, ovate. Initial leaf: green tinged with light red. Flower: white bud tinged with light pink on edge, 5 to 7 flowers per cluster in average of 6.1; stamen number: 17 to 20, averaging 19.4; corolla diameter: 3.8cm. Fruit: matures in late September in Xingcheng, Liaoning Province, 118g per fruit, 5.9cm long, 5.7cm wide, sub-round, greenish-yellow or whitish-yellow skin, thick, medium core, 5 locules, flesh pale yellowish-white, crisp, mid-coarse, juicy, sour-sweet or light sweet; TSS 13.3%, TA 0.11%; quality above medium; storage life long.

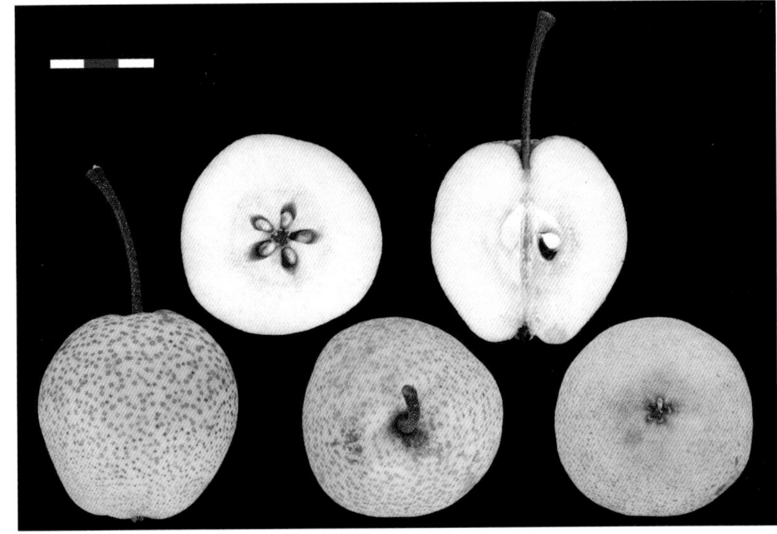

21. 威宁大黄梨

来源及分布 2n=34，原产贵州威宁。

主要性状 树势中庸，丰产性中等，始果年龄中等。叶片长12.7cm，宽7.3cm，卵圆形，初展叶红色，有茸毛。花蕾白色，边缘浅粉红色，每花序4～9朵花，平均6.3朵；雄蕊21~30枚，平均25.2枚。在湖北武汉，果实9月上旬成熟，单果重216g，纵径7.4cm，横径7.3cm，近圆形或倒卵圆形；果皮褐色或绿褐色，厚；果心小，5心室；果肉淡黄白色，肉质中粗，松脆，汁液多，味酸甜；含可溶性固形物13.20%；品质中上等。

21. Weining Dahuangli

Origin and Distribution Weining Dahuangli (2n=34), was originated in Weining, Guizhou Province.

Main Characters Tree: moderately vigorous and productive, medium precocity. Leaf: 12.7cm × 7.3cm in size, ovate. Initial leaf: red, pubescent. Flower: white bud tinged with light pink on edge, 4 to 9 flowers per cluster in average of 6.3; stamen number: 21 to 30, averaging 25.2. Fruit: matures in early September in Wuhan, Hubei Province, 216g per fruit, 7.4cm long, 7.3cm wide, sub-round or obovate, greenish-brown or brown skin, thick, small core, 5 locules, flesh pale yellowish-white, mid-coarse, crisp tender, juicy, sour-sweet; TSS 13.20%; quality above medium.

22. 细花红梨

来源及分布　2n=34，原产广东惠阳。

主要性状　树势强，丰产，始果年龄早。叶片长11.4cm，宽6.2cm，卵圆形，初展叶褐红色。花蕾白色，边缘浅粉红色，每花序2～9朵花，平均6.4朵；雄蕊16～23枚，平均20.1枚。在湖北武汉，果实9月中旬成熟，单果重198g，纵径5.5cm，横径6.4cm，扁圆形；果皮黄褐色，厚，粗糙；果心中大，5心室；果肉白色，肉质细，松脆，汁液多，味酸甜或甜；含可溶性固形物13.4%，可滴定酸0.30%；品质上等或中上等。

22. Xihua Hongli

Origin and Distribution Xihua Hongli (2n=34), was originated in Huiyang, Guangdong Province.

Main Characters Tree: vigorous, productive, early precocity. Leaf: 11.4cm × 6.2cm in size, ovate. Initial leaf: brownish-red. Flower: white bud tinged with light pink on edge, 2 to 9 flowers per cluster in average of 6.4; stamen number: 16 to 23, averaging 20.1. Fruit: matures in mid-September in Wuhan, Hubei Province, 198g per fruit, 5.5cm long, 6.4cm wide, oblate, yellowish-brown skin, thick, rough, medium core, 5 locules, flesh white, fine, crisp tender, juicy, sour-sweet or sweet; TSS 13.4%, TA 0.30%; good or above medium in quality.

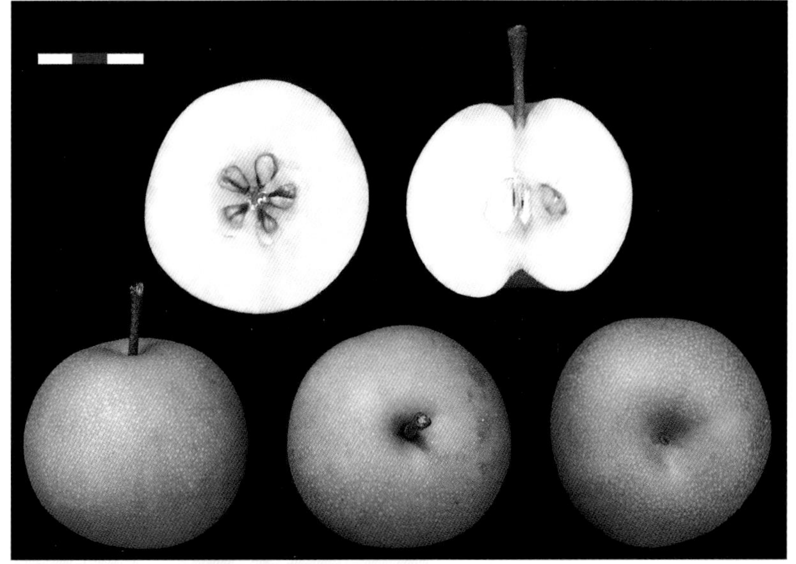

23. 鸭蛋青

来源及分布 2n=34，原产湖南靖县。

主要性状 树势中庸，丰产性中等，始果年龄晚。叶片长 11.1cm，宽 6.9cm，卵圆形，初展叶绿色，着红色。花蕾白色，每花序 4 ~ 6 朵花，平均 4.9 朵；雄蕊 19 ~ 27 枚，平均 22.0 枚。在湖北武汉，果实 9 月上、中旬成熟，单果重 165g，纵径 7.9cm，横径 6.2cm，长圆形或倒卵圆形；果皮绿色或绿黄色；果心中大，5 心室；果肉淡黄白色，肉质较细，脆，汁液中多，味酸甜；含可溶性固形物 10.8%，可滴定酸 0.26%；品质中上等。

23. Yadanqing

Origin and Distribution Yadanqing (2n=34), was originated in Jingxian, Hunan Province.

Main Characters Tree: moderately vigorous and productive, late precocity. Leaf: 11.1cm × 6.9cm in size, ovate. Initial leaf: green tinged with red. Flower: white bud, 4 to 6 flowers per cluster in average of 4.9; stamen number: 19 to 27, averaging 22.0. Fruit: matures in early or mid September in Wuhan, Hubei Province, 165g per fruit, 7.9cm long, 6.2cm wide, long round or obovate, green or greenish-yellow skin, medium core, 5 locules, flesh pale yellowish-white, relatively fine, crisp, mid-juicy, sour-sweet; TSS 10.8%, TA 0.26%; quality above medium.

24. 雁荡雪梨

来源及分布 2n=34，又名大头梨、迟梨、人头梨，原产浙江乐清。

主要性状 树势强，丰产，始果年龄早或中等。叶片长 11.8cm，宽 8.3cm，卵圆形，初展叶红色，微显绿色。花蕾白色，每花序 7 ～ 8 朵花，平均 7.5 朵；雄蕊 20 ～ 28 枚，平均 24.6 枚；花冠直径 3.5cm。在辽宁兴城，果实 9 月下旬至 10 月上旬成熟，单果重 213g，纵径 6.4cm，横径 7.5cm，扁圆形，有些果实形状不规则；果皮绿褐色，厚；果心中大，5 心室；果肉白色，肉质松脆，中粗，汁液多，味甜酸，含可溶性固形物 10.83%，可滴定酸 0.20%；品质中等或中上等。

24. Yandang Xueli

Origin and Distribution Yandang Xueli (2n=34), also known as Datouli, Chili, Rentouli, was originated in Leqing, Zhejiang Province.

Main Characters Tree: vigorous, productive, early or medium precocity. Leaf: 11.8cm × 8.3cm in size, ovate. Initial leaf: red with less green. Flower: white bud, 7 to 8 flowers per cluster in average of 7.5; stamen number: 20 to 28, averaging 24.6; corolla diameter: 3.5cm. Fruit: matures in late September or early October in Xingcheng, Liaoning Province, 213g per fruit, 6.4cm long, 7.5cm wide, oblate, some irregular, greenish-brown skin, thick, medium core, 5 locules, flesh white, crisp tender, mid-coarse, juicy, sour-sweet; TSS 10.83%, TA 0.20%; medium or above in quality.

25. 严州雪梨

来源及分布 2n=34，又名严州白梨、桐庐白梨、白梨，原产浙江，在浙江、四川、福建等地有栽培。

主要性状 树势强，丰产，始果年龄早。叶片长12.5cm，宽8.8cm，卵圆形，初展叶红色。花蕾白色，边缘浅粉红色，每花序5～7朵花，平均6.6朵；雄蕊17～20枚，平均19.0枚；花冠直径3.6cm。在辽宁兴城，果实9月中、下旬成熟，单果重172g，纵径7.2cm，横径6.8cm，倒卵圆形；果皮绿黄色；果心小，5心室；果肉绿白色，肉质细，疏松，汁液多，味甜；含可溶性固形物10.97%，可滴定酸0.08%；品质上等或中上等。

25. Yanzhou Xueli

Origin and Distribution Yanzhou Xueli (2n=34), also known as Yanzhou Baili, Tonglu Baili, or Baili, originated in Zhejiang Province, is grown in Zhejiang, Sichuan and Fujian, etc.

Main Characters Tree: vigorous, productive, early precocity. Leaf: 12.5cm × 8.8cm in size, ovate. Initial leaf: red. Flower: white bud tinged with light pink on edge, 5 to 7 flowers per cluster in average of 6.6; stamen number: 17 to 20, averaging 19.0; corolla diameter: 3.6cm. Fruit: matures in mid or late September in Xingcheng, Liaoning Province, 172g per fruit, 7.2cm long, 6.8cm wide, obovate, greenish-yellow skin, small core, 5 locules, flesh greenish-white, fine, tender, juicy, sweet; TSS 10.97%, TA 0.08%; good or above medium in quality.

26. 早三花

来源及分布 2n=34，又名三花梨，原产浙江义乌。

主要性状 树势中庸，丰产，始果年龄早或中等。叶片长9.9cm，宽6.2cm，卵圆形，初展叶绿色，微着浅红色。花蕾白色，小花蕾边缘浅粉红色，每花序5～6朵花，平均5.6朵；雄蕊20～26枚，平均22.8枚；花冠直径4.3cm。在辽宁兴城，果实9月下旬成熟，单果重139g，纵径6.5cm，横径6.3cm，长圆形；果皮绿黄色或黄色，梗端有锈斑，向萼端锈斑逐渐减少；果心小或中大，5心室；果肉淡黄白色，肉质细，松脆，汁液多，味甜或酸甜；含可溶性固形物13.47%，可滴定酸0.11%；品质上等。

26. Zaosanhua

Origin and Distribution Zaosanhua (2n=34), also known as Sanhuali, was originated in Yiwu, Zhejiang Province.

Main Characters Tree: moderately vigorous, productive, early or medium precocity. Leaf: 9.9cm × 6.2cm in size, ovate. Initial leaf: green tinged with light red. Flower: white bud, small one tinged with light pink on edge, 5 to 6 flowers per cluster in average of 5.6; stamen number: 20 to 26, averaging 22.8; corolla diameter: 4.3cm. Fruit: matures in late September in Xingcheng, Liaoning Province, 139g per fruit, 6.5cm long, 6.3cm wide, long round, greenish-yellow or yellow skin covered with russet on stalk cavity, gradually less russet to basin, small or medium core, 5 locules, flesh pale yellowish-white, fine, crisp tender, juicy, sweet or sour-sweet; TSS 13.47%, TA 0.11%; quality good.

27. 紫酥

来源及分布　原产安徽砀山，在安徽砀山县及周围地区有栽培。

主要性状　树势强，丰产，始果年龄中等。叶片长 11.0cm，宽 6.5cm，阔卵圆形，初展叶红色，带有绿色。花蕾白色，边缘淡粉红色，每花序 5 ～ 7 朵花，平均 5.9 朵；雄蕊 18 ～ 21 枚，平均 19.8 枚；花冠直径 3.4cm。在辽宁兴城，果实 9 月中、下旬成熟，单果重 131g，纵径 6.6cm，横径 6.2cm，近圆形；果皮绿黄色，果面有大面积果锈；果心中大，5 心室；果肉淡黄白色，肉质细，松脆，汁液多，味淡甜；含可溶性固形物 11.55%，可滴定酸 0.10%；品质中上等。

27. Zisu

Origin and Distribution Zisu, originated in Dangshan County, Anhui Province, is grown mainly in Dangshan County and surrounding areas.

Main Characters Tree: vigorous, productive, medium precocity. Leaf: 11.0cm × 6.5cm in size, broadly ovate. Initial leaf: red with green. Flower: white bud tinged with light pink on edge, 5 to 7 flowers per cluster in average of 5.9; stamen number: 18 to 21, averaging 19.8; corolla diameter: 3.4cm. Fruit: matures in mid or late September in Xingcheng, Liaoning Province, 131g per fruit, 6.6cm long, 6.2cm wide, subrounde, greenish-yellow skin covered with obvious russet on surface, medium core, 5 locules, flesh pale yellowish-white, fine, crisp tender, juicy, light sweet; TSS 11.55%, TA 0.10%; quality above medium.

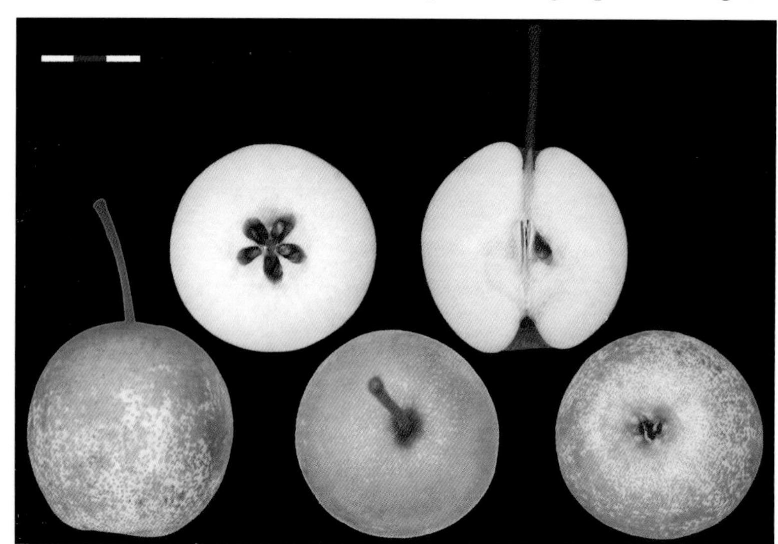

（三）秋子梨品种 Ussurian Pear Varieties

1. 安梨

来源及分布 2n=3x=51，原产东北或河北省燕山山脉地区，在辽宁省北镇、鞍山及河北省兴隆、青龙、抚宁等地栽培较多。

主要性状 树势强，丰产，始果年龄中或晚。叶片长12.0cm，宽9.6cm，近圆形，初展叶黄绿色。花蕾白色，小花蕾边缘浅粉红色，每花序3～5朵花，平均4.4朵；雄蕊19～23枚，平均20.6枚；花冠直径6.1cm。果实10月上旬成熟，单果重106g，纵径5.2cm，横径6.2cm，扁圆形，不规则；果皮较粗糙，黄绿色，厚；果心中大，5个心室；果肉白色或淡黄白色，肉质粗，采收时紧密而脆，味酸稍涩，后熟变软，味甜酸，涩味消失；含可溶性固形物14.23%，可滴定酸1.13%；品质中上等。果实耐贮藏。

1. Anli

Origin and Distribution Anli (2n=3x=51), originated in Northeast China or Yanshan mountains in Hebei Province, is grown widely in Beizhen and Anshan of Liaoning Province, also in Xinglong, Qinglong, and Funing of Hebei Province.

Main Characters Tree: vigorous, productive, medium or late precocity. Leaf: 12.0cm × 9.6cm in size, sub-round. Initial leaf: yellowish-green. Flower: white bud, small one tinged with light pink on edge, 3 to 5 flowers per cluster in average of 4.4; stamen number: 19 to 23, averaging 20.6; corolla diameter: 6.1cm. Fruit: matures in early October, 106g per fruit, 5.2cm long, 6.2cm wide, oblate, irregular, yellowish-green skin, rough, thick, medium core, 5 locules, flesh white or light yellowish-white, coarse, dense and crisp, sour, little astringency when harvested, turning to soft after ripening, sweet-sour ; TSS 14.23%, TA 1.13%; quality above medium; storage life long.

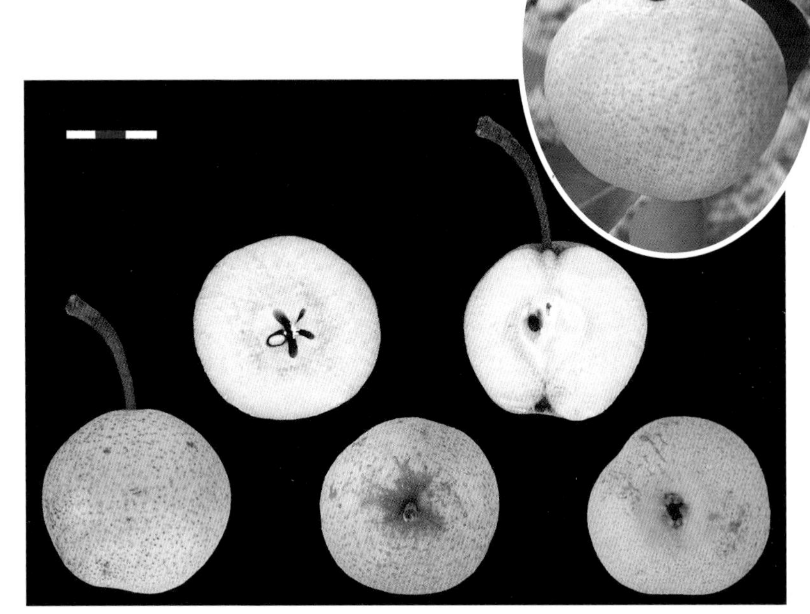

2. 八里香

来源及分布 2n=34，原产辽宁省，在建昌、绥中、兴城、凌海、义县等地有栽培。

主要性状 树势强，丰产，始果年龄中等。叶片长9.7cm，宽6.9cm，阔卵圆形，初展叶黄绿色，微显红色。花蕾白色，边缘粉红色，每花序7～11朵花，平均8.5朵；雄蕊20枚；花冠直径5.0cm。果实9月上、中旬成熟，单果重43g，纵径4.1cm，横径4.2cm，扁圆形；果皮绿黄色，阳面着淡红晕；果心大，5心室；果肉黄白色，肉质粗，采收时肉质紧密而韧，酸涩，10d左右后熟，汁液中多，肉质变软，味甜酸，涩味消失，香气浓；含可溶性固形物13.17%，可滴定酸1.0%；品质中上等。果实不耐贮藏。

2. Balixiang

Origin and Distribution Balixiang (2n=34), originated in Liaoning Province, is grown mainly in Jianchang, Suizhong, Xingcheng, Linghai and Yixian, etc.

Main Characters Tree: vigorous, productive, medium precocity. Leaf: 9.7cm × 6.9cm in size, broadly ovate. Initial leaf: yellowish-green tinged with light red. Flower: white bud tinged with pink on edge, 7 to 11 flowers per cluster in average of 8.5; stamen number: 20; corolla diameter: 5.0cm. Fruit: matures in early or mid September, 43g per fruit, 4.1cm long, 4.2cm wide, oblate, greenish-yellow skin covered with light red on the side exposed to the sun, large core, 5 locules, flesh yellowish-white, coarse, dense and tough, sour with astringency, turning to soft after 10 days of storage, mid-juicy, sweet-sour without astringency, strongly aromatic; TSS 13.17%, TA 1.0%; quality above medium; storage life short.

3. 大香水

来源及分布 2n=34，又名大头黄，原产辽宁省，在辽宁鞍山栽培较多，吉林延边亦有栽培。

主要性状 树势强，丰产，始果年龄晚。叶片长 10.0cm，宽 7.9cm，阔卵圆形，初展叶绿色，微显红色。花蕾白色，小花蕾边缘浅粉红色，每花序 7 ~ 9 朵花，平均 7.9 朵；雄蕊 20 枚；花冠直径 4.9cm。果实 8 月中、下旬成熟，单果重 101g，纵径 5.3cm，横径 5.7cm，长圆形或阔圆锥形；果皮绿黄色或黄色，少数果实有淡红晕；果心大，5 心室；果肉淡黄白色，肉质中粗，采收时紧密而韧，采后 8d 左右后熟，肉质变软，汁液中多，味甜酸，稍涩，有香气；含可溶性固形物 12.10%，可滴定酸 0.43%；品质中等或中上等。

3. Daxiangshui

Origin and Distribution Daxiangshui (2n=34), also known as Datouhuang, originated in Liaoning Province, is grown mainly in Anshan of Liaoning Province, also cultivated in Yanbian of Jilin Province.

Main Characters Tree: vigorous, productive, late precocity. Leaf: 10.0cm × 7.9cm in size, broadly ovate. Initial leaf: green tinged with light red. Flower: white bud, small one tinged with light pink on edge, 7 to 9 flowers per cluster in average of 7.9; stamen number: 20; corolla diameter: 4.9cm. Fruit: matures in mid or late August, 101g per fruit, 5.3cm long, 5.7cm wide, long round or broadly conical, greenish-yellow or yellow skin, few covered with light red, large core, 5 locules, flesh pale yellowish-white, mid-coarse, dense and tough when harvested, turning to soft after 8 days of storage, mid-juicy, sweet-sour with a little astringency, aromatic; TSS 12.10%, TA 0.43%; medium or above in quality.

4. 花盖

来源及分布　2n=34，原产辽宁省，在辽宁西部、吉林延边及河北燕山山脉地区均栽培较多。

主要性状　树势中庸，丰产，始果年龄中或晚。叶片长 8.8cm，宽 6.8cm，卵圆形，初展叶绿色，着红色。花蕾白色，边缘粉红色，每花序 6 ~ 8 朵花，平均 6.7 朵；雄蕊 19 ~ 21 枚，平均 19.6 枚；花冠直径 4.1cm。果实 9 月下旬至 10 月上旬成熟，单果重 77.5g，纵径 4.5cm，横径 5.4cm，扁圆形；果皮绿黄色或黄色，梗端有片锈；果心中大，4 或 5 心室；果肉淡黄白色，肉质中粗，采收时肉质紧密而硬，后熟果肉变软，汁液多，味甜酸，有香气；含可溶性固形物 15.40%，可滴定酸 0.69%；品质中上等。果实耐贮藏。

4. Huagai

Origin and Distribution Huagai (2n=34), originated in Liaoning Province, is grown mainly in western Liaoning, also in Yanbian of Jilin Province, and Yanshan mountains in Hebei Province.

Main Characters Tree: moderately vigorous, productive, medium or late precocity. Leaf: 8.8cm × 6.8cm in size, ovate. Initial leaf: green tinged with red. Flower: white bud tinged with pink on edge, 6 to 8 flowers per cluster in average of 6.7; stamen number: 19 to 21, averaging 19.6; corolla diameter: 4.1cm. Fruit: matures in late September or early October, 77.5g per fruit, 4.5cm long, 5.4cm wide, oblate, greenish-yellow or yellow skin covered with russet on stalk end, medium core, 4 or 5 locules, flesh pale yellowish-white, mid-coarse, dense and tough when harvested, turning to soft after storage, juicy, sweet-sour, aromatic; TSS 15.40%, TA 0.69%; quality above medium, storage life long.

5. 尖把梨

来源及分布 2n=34，原产辽宁省，在辽宁开原和吉林延边栽培较多。

主要性状 树势强，丰产，始果年龄晚。叶片长12.0cm，宽6.6cm，卵圆形，初展叶黄绿色，着很浅的红色。花蕾白色，边缘浅粉红色，每花序5～8朵花，平均6.7朵；雄蕊20枚；花冠直径4.0cm。果实9月中至下旬成熟，单果重87g，纵径5.3cm，横径5.5cm，短葫芦形；果皮绿黄色或黄色，梗端有片锈；果心中大，4或5心室；果肉淡黄白色，采收时肉质紧密而韧，不宜食用，经后熟，果肉变软，汁液多，味甜酸，味浓，有香气；含可溶性固形物15.36%，可滴定酸1.31%；品质中上等。果实耐贮藏，适宜冻藏。

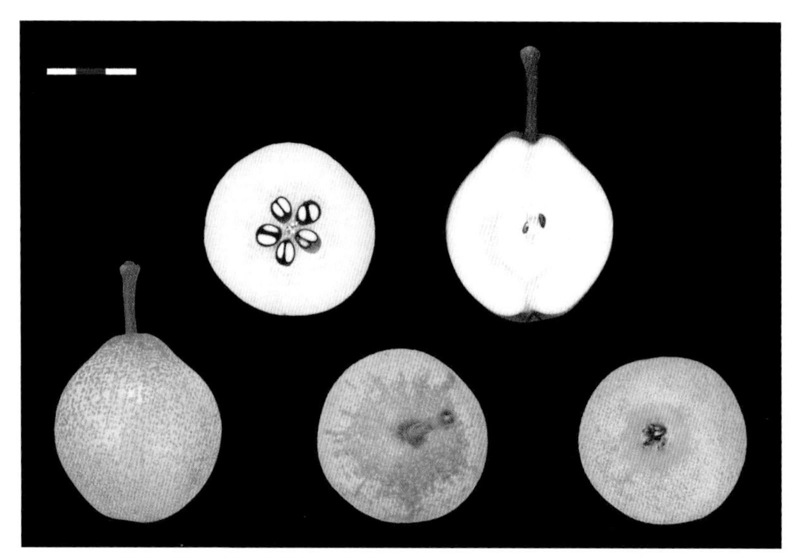

5. Jianbali

Origin and Distribution Jianbali (2n=34), originated in Liaoning Province, is grown mainly in Kaiyuan of Liaoning Province, also in Yanbian of Jilin Province.

Main Characters Tree: vigorous, productive, late precocity. Leaf: 12.0cm × 6.6cm in size, ovate. Initial leaf: yellowish-green tinged with very light red. Flower: white bud tinged with light pink on edge, 5 to 8 flowers per cluster in average of 6.7; stamen number: 20; corolla diameter: 4.0cm. Fruit: matures in mid or late September, 87g per fruit, 5.3cm long, 5.5cm wide, short pyriform, greenish-yellow or yellow skin covered with russet on stalk end, medium core, 4 or 5 locules, flesh light yellowish white, dense and rough when harvested, not suitable for eating, turning to soft after storage, juicy, strongly sweet-sour, aromatic; TSS 15.36%, TA 1.31%; quality above medium; storage life long, suitable for frozen eating.

6. 京白梨

来源及分布 2n=34，原产北京，在北京近郊、东北南部、山西、西北各地有栽培。

主要性状 树势中庸，丰产，始果年龄晚。叶片长 8.7cm，宽 6.8cm，卵圆形，初展叶绿黄色。花蕾白色，每花序 6 ～ 9 朵花，平均 7.6 朵；雄蕊 19 ～ 21 枚，平均 19.9 枚；花冠直径 3.7cm。果实 9 月上、中旬成熟，单果重 121g，纵径 5.2cm，横径 6.3cm，扁圆形；果皮黄绿色或黄白色；果心中大，5 或 6 心室；果肉黄白色，采收时肉质脆，10 ～ 14d 后熟，肉质变软，中粗或细，汁液多，味甜，微有香气；含可溶性固形物 14.83%，可滴定酸 0.82%；品质上等。果实不耐贮藏。

6. Jingbaili

Origin and Distribution Jingbaili (2n=34), originated in Beijing, is grown mainly in suburbs of Beijing, southern parts of northest China, also in Shanxi and northwest China.

Main Characters Tree: moderately vigorous, productive, late precocity. Leaf: 8.7cm × 6.8cm in size, ovate. Initial leaf: greenish-yellow. Flower: white bud, 6 to 9 flowers per cluster in average of 7.6; stamen number: 19 to 21, averaging 19.9; corolla diameter: 3.7cm. Fruit: matures in early or mid September, 121g per fruit, 5.2cm long, 6.3cm wide, oblate, yellowish-green or whitish-yellow skin, medium core, 5 or 6 locules, flesh yellowish-white, crisp when harvested, turning to soft after 10 to 14 days of storage, mid-coarse or fine, juicy, sweet, a little aromatic; TSS 14.83%, TA 0.82%; quality good; storage life short.

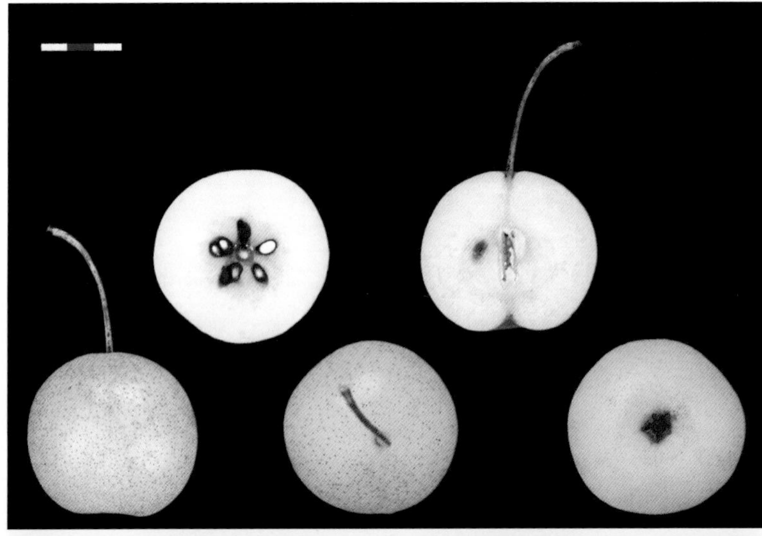

7. 满园香

来源及分布 2n=34，原产辽宁省，在绥中、兴城、北镇等地有栽培。

主要性状 树势强，产量中等，始果年龄晚。叶片长 10.8cm，宽 7.0cm，卵圆形，初展叶绿黄色。花蕾白色，小花蕾边缘浅粉红色，每花序 6 ~ 8 朵花，平均 7.1 朵；雄蕊 18 ~ 20 枚，平均 19.4 枚；花冠直径 3.6cm。果实 8 月下旬至 9 月上旬成熟，单果重 49g，纵径 4.0cm，横径 4.6cm，扁圆形，有些果实梗端稍尖；果皮绿黄色或黄白色；果心大，5 心室；果肉黄白色，采收时紧密而脆，稍涩，经 7d 左右后熟变软，汁液多，甜酸味浓，有香气；含可溶性固形物 15.67%，可滴定酸 0.86%；品质中上等。果实不耐贮藏。

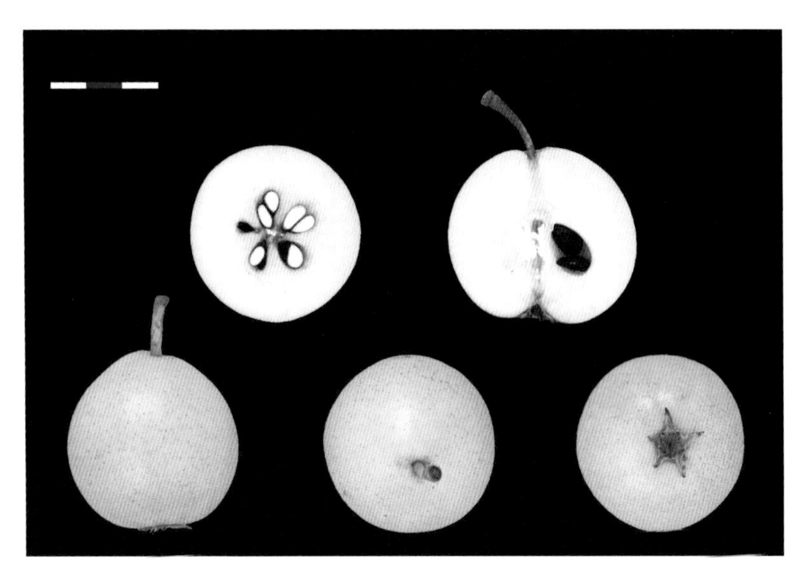

7. Manyuanxiang

Origin and Distribution Manyuanxiang (2n=34), originated in Liaoning Province, is grown mainly in Suizhong, Xingcheng, and Beizhen, etc.

Main Characters Tree: vigorous, medium productive, late precocity. Leaf: 10.8cm × 7.0cm in size, ovate. Initial leaf: greenish-yellow. Flower: white bud, small one tinged with light pink on edge, 6 to 8 flowers per cluster in average of 7.1; stamen number: 18 to 20, averaging 19.4; corolla diameter: 3.6cm. Fruit: matures in late August or early September, 49g per fruit, 4.0cm long, 4.6cm wide, oblate, some subacute at stalk end, greenish-yellow or whitish-yellow skin, large core, 5 locules, flesh yellowish-white, dense and crisp when harvested, a little astringent, turning to soft after 7 days of storage, juicy, strongly sweet-sour, aromatic; TSS 15.67%, TA 0.86%; quality above medium; storage life short.

8. 南果梨

来源及分布　2n=34，原产辽宁省鞍山市，在鞍山、辽阳等地栽培最多，在吉林、内蒙古、山西等地亦有栽培。

主要性状　树势中庸，丰产，产量不稳定，始果年龄中等。叶片长 9.6cm，宽 5.9cm，卵圆形，初展叶黄绿色。花蕾白色，边缘粉红色，每花序 6 ~ 9 朵花，平均 7.4 朵；雄蕊 20 ~ 26 枚，平均 22.5 枚；花冠直径 3.8cm。果实 9 月上、中旬成熟，单果重 58g，纵径 4.2cm，横径 4.8cm，圆形或扁圆形；果皮绿黄色或黄色，有些果实有红晕；果心中大，4 或 5 心室；果肉淡黄白色，采收时肉质脆，经 15 ~ 20d 后熟，肉质软溶，汁液多，甜酸味浓，浓香；含可溶性固形物 15.50%，可滴定酸 0.56%；品质极上。

8. Nanguoli

Origin and Distribution Nanguoli (2n=34), originated in Anshan, Liaoning Province, is widely grown in Anshan and Liaoyang of Liaoning Province, also grown in Jilin, Inner Mongolia, Shanxi Provinces, etc.

Main Characters Tree: moderately vigorous, productive but unstable, medium precocity. Leaf: 9.6cm × 5.9cm in size, ovate. Initial leaf: yellowish-green. Flower: white bud tinged with pink on edge, 6 to 9 flowers per cluster in average of 7.4; stamen number: 20 to 26, averaging 22.5; corolla diameter: 3.8cm. Fruit: matures in early or mid September, 58g per fruit, 4.2cm long, 4.8cm wide, oblate or globose, greenish-yellow or yellow skin, some covered with red blush, medium core, 4 or 5 locules, flesh light yellowish-white, crisp when harvested, turning to melting after 15 to 20 days of storage, juicy, strongly sweet-sour, aromatic; TSS 15.50%, TA 0.56%; extremely good in quality.

9. 秋子

来源及分布 2n=34，原产辽宁省，在北镇、义县等地栽培较多。

主要性状 树势强，丰产，始果年龄晚。叶片长10.8cm，宽7.7cm，卵圆形，初展叶红色，带绿色。花蕾白色，边缘粉红色，部分花瓣边缘亦呈粉红色，每花序5～8朵花，平均6.8朵；雄蕊19～20枚，平均19.5枚；花冠直径3.0cm。果实9月下旬成熟，单果重55g，纵径4.5cm，横径4.8cm，近圆形，萼端略细；果皮绿黄色或黄白色，有些果实阳面有淡红色晕；果心大，5心室；果肉白色，肉质中粗，脆，汁液多，味酸或甜酸，微香；含可溶性固形物12.58%，可滴定酸0.83%；品质中上等。

9. Qiuzi

Origin and Distribution Qiuzi (2n=34), originated in Liaoning Province, is grown mainly in Beizhen and Yixian, etc.

Main Characters Tree: vigorous, productive, late precocity. Leaf: 10.8cm × 7.7cm in size, ovate. Initial leaf: red with green. Flower: white bud tinged with pink on edge, some petals also pink on edge, 5 to 8 flowers per cluster in average of 6.8; stamen number: 19 to 20, averaging 19.5; corolla diameter: 3.0cm. Fruit: matures in late September, 55g per fruit, 4.5cm long, 4.8cm wide, subglobose, somewhat thin on cavity end, greenish-yellow or whitish-yellow skin, some covered with light red on the side exposed to the sun, large core, 5 locules, flesh white, mid-coarse, crisp, juicy, sour or sweet-sour, slightly aromatic; TSS 12.58%, TA 0.83%; quality above medium.

10. 甜秋子

来源及分布 2n=34，原产辽宁省，在北镇有少量栽培。

主要性状 树势中庸，丰产，始果年龄中或晚。叶片长 10.4cm，宽 8.0cm，卵圆形，初展叶红色，带有绿色。花蕾白色，边缘浅粉红色，每花序 6 ~ 8 朵花，平均 7.0 朵；雄蕊 20 枚；花冠直径 3.6cm。果实 9 月下旬成熟，单果重 101g，纵径 5.1cm，横径 5.8cm，扁圆形；果皮绿黄色或黄白色，部分果实着浅红色；果心中大或大，5 心室；果肉白色，肉质较细，松脆，汁液多，味酸甜；含可溶性固形物 12.55%，可滴定酸 0.44%；品质中上或上等。

10. Tianqiuzi

Origin and Distribution Tianqiuzi (2n=34), originated in Liaoning Province, is grown in Beizhen of Liaoning Province limitedly.

Main Characters Tree: moderately vigorous, productive, medium or late precocity. Leaf: 10.4cm × 8.0cm in size, ovate. Initial leaf: red with green. Flower: white bud tinged with light pink on edge, 6 to 8 flowers per cluster in average of 7.0; stamen number: 20; corolla diameter: 3.6cm. Fruit: matures in late September, 101g per fruit, 5.1cm long, 5.8cm wide, oblate, greenish-yellow or whitish-yellow skin, some covered with pale red, medium or large core, 5 locules, flesh white, relatively fine, crisp tender, juicy, sour-sweet; TSS 12.55%, TA 0.44%; above medium or good in quality.

11. 软儿梨

来源及分布 2n=34，原产甘肃省，在青海、甘肃、宁夏等地有栽培。

主要性状 树势强，丰产，始果年龄中等。叶片长8.7cm，宽6.3cm，卵圆形，初展叶绿色，微显红色。花蕾白色，每花序6～8朵花，平均6.8朵；雄蕊19～23枚，平均20.4枚；花冠直径4.3cm。果实9月下旬成熟，单果重61g，纵径4.4cm，横径5.0cm，扁圆形，果面略显不平；果皮黄绿色或黄色；果心中大，5心室；果肉淡黄白色，肉质粗，紧密而韧，汁液少，贮藏1个月左右后熟，果肉变软，味甜酸，有香气；含可溶性固形物13.43%，可滴定酸0.74%；品质中等。冻藏肉质软，汁液多，较受欢迎。

11. Ruan'erli

Origin and Distribution Ruan'erli (2n=34), originated in Gansu Province, is grown in Qinghai, Gansu, and Ningxia, etc.

Main Characters Tree: vigorous, productive, medium precocity. Leaf: 8.7cm × 6.3cm in size, ovate. Initial leaf: green slightly tinged with red. Flower: white bud, 6 to 8 flowers per cluster in average of 6.8; stamen number: 19 to 23, averaging 20.4; corolla diameter: 4.3cm. Fruit: matures in late September, 61g per fruit, 4.4cm long, 5.0cm wide, oblate, a little rough, yellowish-green or yellow skin, medium core, 5 locules, flesh light yellowish-white, coarse, dense and tough, less juicy, turning to soft after 1 month of storage, sweet-sour, aromatic; TSS 13.43%, TA 0.74%; quality medium. Fruits become juicy after frozen storage and loved by consumers.

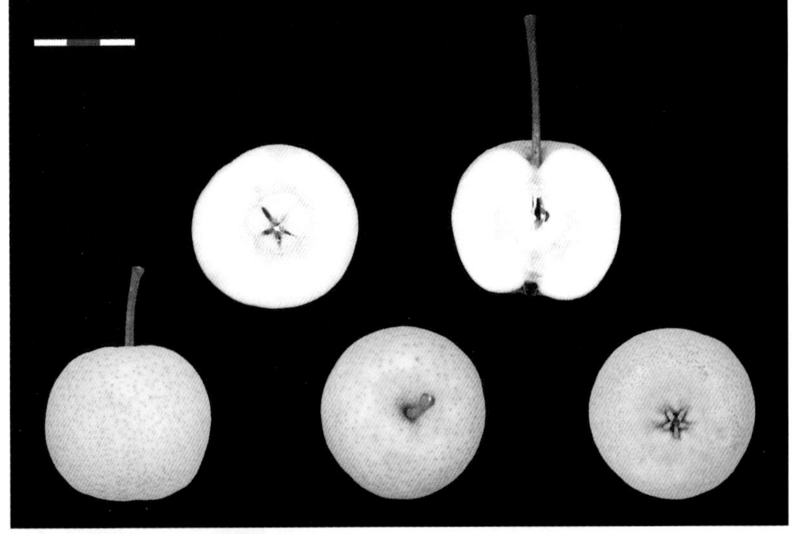

12. 五香梨

来源及分布 2n=34，又名五节香、水五香、臭盖子、羊五香，原产辽宁省北镇市，在辽宁北镇、河北抚宁和青龙等地有栽培。

主要性状 树势强，产量中等，始果年龄晚。叶片长12.1cm，宽7.1cm，阔卵圆形，初展叶红色，微带绿色。花蕾白色，每花序6～9朵花，平均7.5朵；雄蕊20～21枚，平均20.3枚；花冠直径5.3cm。果实9月中、下旬成熟，单果重118g，纵径5.5cm，横径6.2cm，扁圆形或阔圆锥形；果皮黄绿色，阳面有红晕，果皮厚；果心中大，5心室；果肉黄白色，采收时肉质紧密而韧，味酸，稍涩，10d左右后熟，肉质变软，涩味消失，味甜酸，香气浓；含可溶性固形物13.60%，可滴定酸0.81%；品质中上等。果实不耐贮藏。

12. Wuxiangli

Origin and Distribution Wuxiangli (2n=34), also known as Wujiexiang, Shuiwuxiang, Chougaizi, Yangwuxiang, originated in Beizhen, Liaoning Province, is grown in Beizhen of Liaoning Province, also in Funing and Qinglong of Hebei Province.

Main Characters Tree: vigorous, medium productive, late precocity. Leaf: 12.1cm × 7.1cm in size, broadly ovate. Initial leaf: red with less green. Flower: white bud, 6 to 9 flowers per cluster in average of 7.5; stamen number: 20 to 21, averaging 20.3; corolla diameter: 5.3cm. Fruit: matures in mid or late September, 118g per fruit, 5.5cm long, 6.2cm wide, oblate or broadly conical, yellowish-green skin covered with red blush on sun side, thick, medium core, 5 locules, flesh yellowish-white, dense and tough, sour with astringency when harvested, turning to soft after 10 days of storage, sweet-sour without astringency, strongly aromatic; TSS 13.60%, TA 0.81%; quality above medium; storage life short.

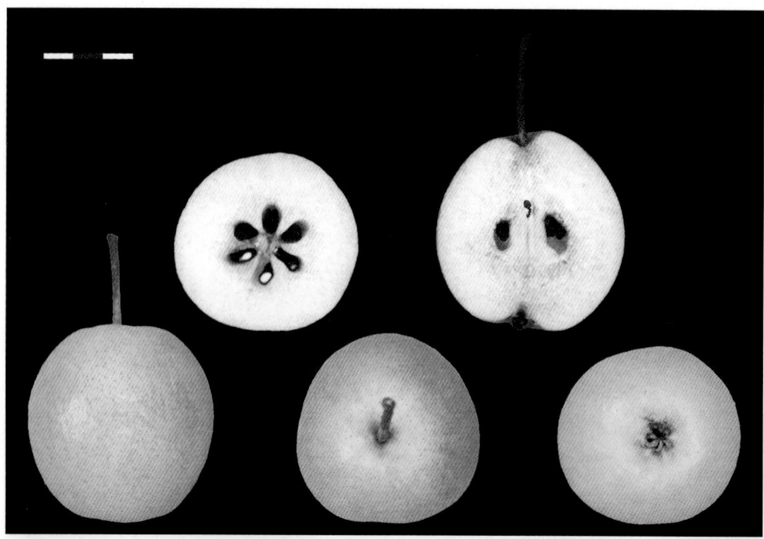

13. 香水梨

来源及分布 2n=34，别名小香水，原产东北，分布较广，尤以辽宁北镇地区栽培较多。

主要性状 树势强，丰产，始果年龄晚。叶片长10.6cm，宽6.1cm，卵圆形，初展叶绿色，着红色。花蕾白色或边缘浅粉红色，每花序6～8朵花，平均7.1朵；雄蕊17～21枚，平均19.1枚；花冠直径4.2cm。果实8月下旬或9月上旬成熟，单果重43g，纵径4.3cm，横径4.2cm，圆形或长圆形；果皮绿黄色或黄色；果心中大，5心室；果肉浅黄白色，肉质较细，采收时松脆，9d左右后熟，肉质变软溶，汁液多，味甜酸，具浓香；含可溶性固形物12.10%，可滴定酸0.80%；品质中上等。果实不耐贮藏。

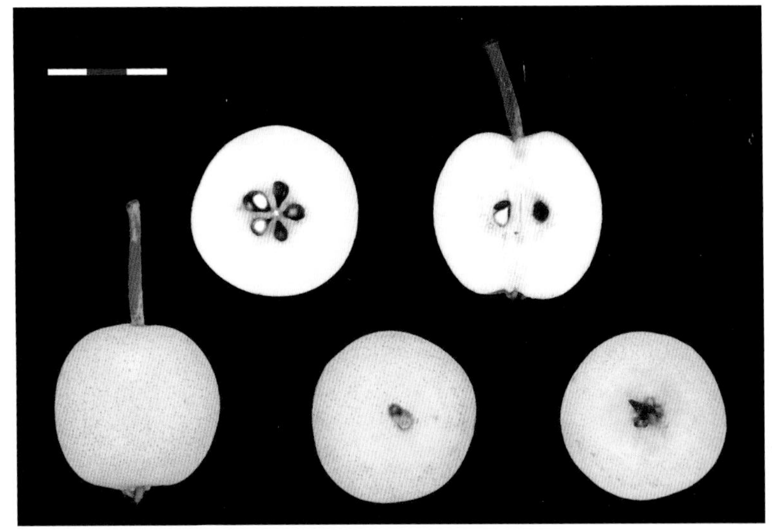

13. Xiangshuili

Origin and Distribution Xiangshuili (2n=34), also known as Xiaoxiangshui, originated in Northeast China, is grown widely there, especially in Beizhen of Liaoning Province.

Main Characters Tree: vigorous, productive, late precocity. Leaf: 10.6cm × 6.1cm in size, ovate. Initial leaf: green tinged with red. Flower: white bud tinged with light pink on edge, 6 to 8 flowers per cluster in average of 7.1; stamen number: 17 to 21, averaging 19.1; corolla diameter: 4.2cm. Fruit: matures in late August or early September, 43g per fruit, 4.3cm long, 4.2cm wide, globose or long globose, greenish-yellow or yellow skin, medium core, 5 locules, flesh pale yellowish-white, relatively fine, crisp tender when harvested, turning to melting after 9 days of storage, juicy, sweet-sour, strongly aromatic; TSS 12.10%, TA 0.80%; quality above medium; storage life short.

14. 鸭广梨

来源及分布 2n=34，原产北京或河北省东北部，在北京近郊、天津武清县、河北东北部栽培较多。

主要性状 树势强，丰产，始果年龄晚。叶片长10.4cm，宽5.9cm，卵圆形，初展叶黄绿色，微着红色。花蕾白色，边缘浅粉红色，每花序8 ~ 10朵花，平均8.5朵；雄蕊20枚；花冠直径5.6cm。果实9月中旬成熟，单果重87g，纵径5.2cm，横径5.4cm，倒卵圆形或扁圆形，不规则，果面凹凸不平。果皮绿色、黄绿色或黄色；果心中大，5心室；果肉淡黄白色，肉质中粗，采收时硬，汁液较少，经8 ~ 9d后熟，肉质变软，汁液增多，味甜酸而浓，具香气；含可溶性固形物13.05%；品质中上等。

14. Yaguangli

Origin and Distribution Yaguangli (2n=34), originated in Beijing or northeast Hebei Province, is grown mainly in suburbs of Beijing, also in Wuqing County of Tianjin, and northeast Hebei Province.

Main Characters Tree: vigorous, productive, late precocity. Leaf: 10.4cm × 5.9cm in size, ovate. Initial leaf: yellowish-green slightly tinged with red. Flower: white bud tinged with light pink on edge, 8 to 10 flowers per cluster in average of 8.5; stamen number: 20; corolla diameter: 5.6cm. Fruit: matures in mid-September, 87g per fruit, 5.2cm long, 5.4cm wide, obovate or oblate, irregular, rough; green, or yellowish-green, or yellow skin, medium core, 5 locules, flesh pale yellowish-white, mid coarse, tough when harvested, less juicy, turning to soft and juicy after 8 to 9 days of storage, strongly sweet-sour, aromatic; TSS 13.05%; quality above medium.

（四）新疆梨品种 Sinkiang Pear Varieties

1. 贵德甜梨

来源及分布 2n=34，又名贵德长把梨、鸡大腿梨，原产青海。在青海省的黄河流域，甘肃的临夏、临洮一带有栽培。

主要性状 树势强，丰产。叶片长 8.5cm，宽 6.0cm，卵圆形或近圆形，初展叶绿色，微着红色。花蕾白色，边缘粉红色，每花序 5 ~ 8 朵花，平均 6.5 朵；雄蕊 19 ~ 21 枚，平均 19.9 枚；花冠直径 4.2cm。在辽宁兴城，果实 8 月中、下旬成熟，单果重 74g，纵径 5.7cm，横径 5.3cm，短葫芦形；果皮绿黄色；果心中大，5 心室；果肉白色，肉质松脆，中粗，汁液中多，味酸甜；含可溶性固形物 12.83%，可滴定酸 0.36%；品质中上等。

1. Guide Tianli

Origin and Distribution Guide Tianli (2n=34), also known as Guide Changbali, or Jidatuili, originated in Qinghai, is grown along Yellow River basin in Qinghai Province, also in Linxia and Lintao of Gansu Province.

Main Characters Tree: vigorous, productive. Leaf: 8.5cm × 6.0cm in size, ovate or sub-round. Initial leaf: green tinged with light red. Flower: white bud tinged with pink on edge, 5 to 8 flowers per cluster in average of 6.5; stamen number: 19 to 21, averaging 19.9; corolla diameter: 4.2cm. Fruit: matures in mid or late August in Xingcheng, Liaoning Province, 74g per fruit, 5.7cm long, 5.3cm wide, short pyriform, greenish-yellow skin, medium core, 5 locules, flesh white, crisp, mid-coarse, mid-juicy, sour-sweet; TSS 12.83%, TA 0.36%; quality above medium.

2. 库尔勒香梨

来源及分布 2n=34，原产新疆，在新疆南部栽培最多，以库尔勒地区生产的最为有名。

主要性状 树势强，丰产，始果年龄中等。叶片长11.9cm，宽5.6cm，卵圆形，初展叶绿色，微显红色。花蕾白色，边缘粉红色，每花序6～7朵花，平均6.5朵；雄蕊20～29枚，平均25.2枚；花冠直径3.8cm。在新疆库尔勒，果实9月上旬成熟，单果重107g，纵径5.6cm，横径6.4cm，纺锤形；果皮绿黄色，阳面有条红，果皮薄；果心大，5心室；果肉白色，肉质细，松脆，汁液多，味甜，有香气；含可溶性固形物14.99%，可滴定酸0.07%；品质上等。果实耐贮藏。

2. Korla Pear

Origin and Distribution Korla Pear (2n=34), originated in Xinjiang, is mostly grown in southern Xinjiang. Fruits produced in Korla are the most famous.

Main Characters Tree: vigorous, productive, medium precocity. Leaf: 11.9cm × 5.6cm in size, ovate. Initial leaf: green tinged with light red. Flower: white bud tinged with pink on edge, 6 to 7 flowers per cluster in average of 6.5; stamen number: 20 to 29, averaging 25.2; corolla diameter: 3.8cm. Fruit: matures in early September in Korla, Xinjiang, 107g per fruit, 5.6cm long, 6.4cm wide, spindle-shaped, greenish-yellow skin covered with striped red, thin, large core, 5 locules, flesh white, fine, juicy, crisp tender, sweet, aromatic; TSS 14.99%, TA 0.07%; quality good; storage life long.

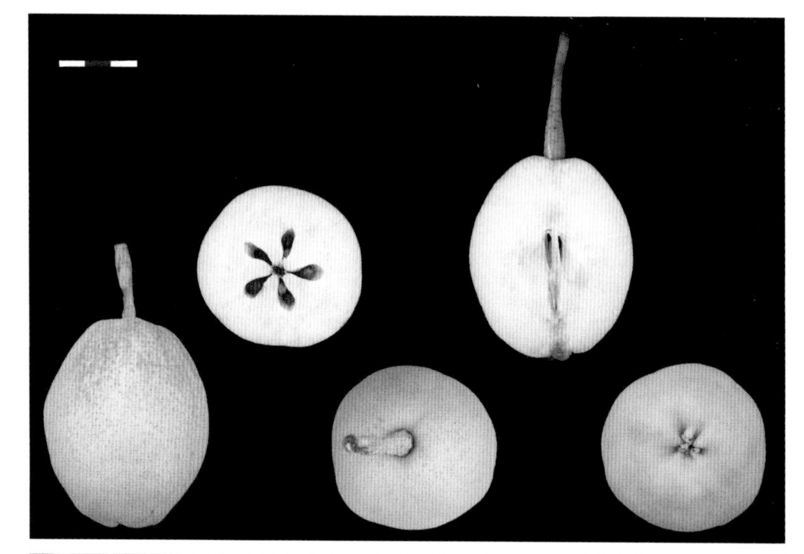

3. 奎克句句

来源及分布 2n=34，别名绿句句，原产新疆。

主要性状 树势较弱，产量中等，始果年龄晚。叶片长10.4cm，宽7.4cm，卵圆形，初展叶绿色。花蕾白色，每花序6～8朵花，平均7.6朵；雄蕊17～25枚，平均22.0枚；花冠直径3.5cm。在辽宁兴城，果实9月上、中旬成熟，单果重62g，纵径4.5cm，横径4.9cm，圆形；果皮黄绿色，有片锈；果心中大，5心室；果肉淡黄白色，肉质细，疏松，汁液多，味酸甜或甘甜；含可溶性固形物17.06%，可滴定酸0.35%；品质中上等。

3. Kuike Juju

Origin and Distribution Kuike Juju (2n=34), also known as Lüjuju, was originated in Xinjiang.

Main Characters Tree: relatively weak, medium productive, late precocity. Leaf: 10.4cm × 7.4cm in size, ovate. Initial leaf: green. Flower: white bud, 6 to 8 flowers per cluster in average of 7.6; stamen number: 17 to 25, averaging 22.0; corolla diameter: 3.5cm. Fruit: matures in early or mid September, 62g per fruit, 4.5cm long, 4.9cm wide, round, yellowish-green skin covered with large russet, medium core, 5 locules, flesh pale yellowish-white, tender, fine, juicy, sour-sweet or very sweet; TSS 17.06%, TA 0.35%; quality above medium.

4. 兰州长把梨

来源及分布 2n=34，原产甘肃省，主要分布在甘肃兰州、张掖等地。

主要性状 树势较弱，较丰产，始果年龄早或中。叶片长7.2cm，宽4.8cm，卵圆形，初展叶绿色，着红色。花蕾白色，边缘浅粉红色，每花序6～8朵花，平均7.4朵；雄蕊19～20枚，平均19.4枚；花冠直径3.5cm。在辽宁兴城，果实8月下旬或9月上旬成熟，单果重80g，纵径6.0cm，横径5.1cm，短葫芦形；果皮绿黄色；果心中大，5心室；果肉淡黄白色，肉质中粗，疏松，汁液中多，味甜酸；含可溶性固形物13.52%，可滴定酸0.57%；品质中上等。

4. Lanzhou Changbali

Origin and Distribution Lanzhou Changbali (2n=34), originated in Gansu Province, is grown mainly in Lanzhou, Zhangye, etc.

Main Characters Tree: relatively weak, relatively productive, early or medium precocity. Leaf: 7.2cm × 4.8cm in size, ovate. Initial leaf: green tinged with red. Flower: white bud tinged with very light pink on edge, 6 to 8 flowers per cluster in average of 7.4; stamen number: 19 to 20, averaging 19.4; corolla diameter: 3.5cm. Fruit: matures in late August or early September in Xingcheng, Liaoning Province, 80g per fruit, 6.0cm long, 5.1cm wide, short pyriform, greenish-yellow skin, medium core, 5 locules, flesh pale yellowish-white, mid-coarse, mid-juicy, tender, sweet-sour; TSS 13.52%, TA 0.57%; quality above medium.

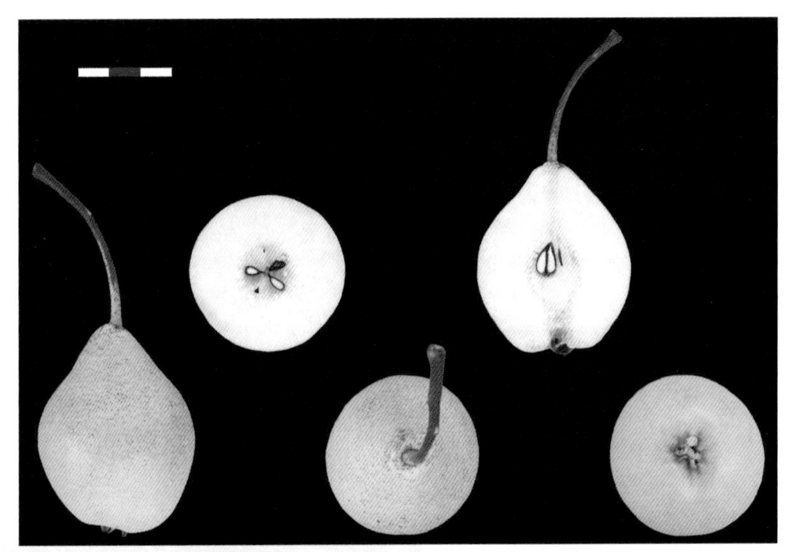

5. 兰州花长把

来源及分布　2n=34，原产甘肃兰州。

主要性状　树势较弱，产量中等，始果年龄早或中。叶片长 6.8cm，宽 5.1cm，卵圆形，初展叶绿色，微显红色。花蕾白色，边缘粉红色，每花序 6 ~ 8 朵花，平均 7.4 朵；雄蕊 19 ~ 21 枚，平均 19.6 枚；花冠直径 3.6cm。在辽宁兴城，果实 8 月下旬或 9 月上旬成熟，单果重 72g，纵径 5.7cm，横径 5.1cm，短葫芦形；果皮绿黄色，果面有黄绿相间的纵向条纹，不规则；果心中大，5 心室；果肉淡黄白色，肉质中粗，疏松，汁液多，味甜酸或甜；含可溶性固形物 11.80%，可滴定酸 0.28%；品质中上等或中等。

5. Lanzhou Huachangba

Origin and Distribution　Lanzhou Huachangba (2n=34), was originated in Lanzhou, Gansu Province.

Main Characters　Tree: relatively weak, medium productive, early or medium precocity. Leaf: 6.8cm × 5.1cm in size, ovate. Initial leaf: green tinged with light red. Flower: white bud tinged with pink on edge, 6 to 8 flowers per cluster in average of 7.4; stamen number: 19 to 21, averaging 19.6; corolla diameter: 3.6cm. Fruit: matures in late August or early September in Xingcheng, Liaoning Province, 72g per fruit, 5.7cm long, 5.1cm wide, short pyriform, greenish-yellow skin covered with longitudinal stripes, irregular; medium core, 5 locules, flesh pale yellowish-white, mid-coarse, tender, juicy, sweet-sour or sweet; TSS 11.80%, TA 0.28%; quality medium or above.

6. 色尔克甫

来源及分布　2n=34，原产新疆。

主要性状　树势中庸或弱，丰产性中等。叶片长8.0cm，宽4.7cm，椭圆形，初展叶绿色，微有很浅的红色。花蕾白色，每花序6～9朵花，平均7.4朵；雄蕊20～24枚，平均21.5枚；花冠直径3.3cm。在辽宁兴城，果实9月中旬成熟，单果重64g，纵径4.9cm，横径4.9cm，近圆形或阔倒卵圆形；果皮绿黄色或黄色，部分果实阳面有淡红晕；果心中大，5心室；果肉白色，肉质疏松，较细，汁液多或中多，味酸甜或甘甜；含可溶性固形物13.43%，可滴定酸0.13%；品质中上等或中等。

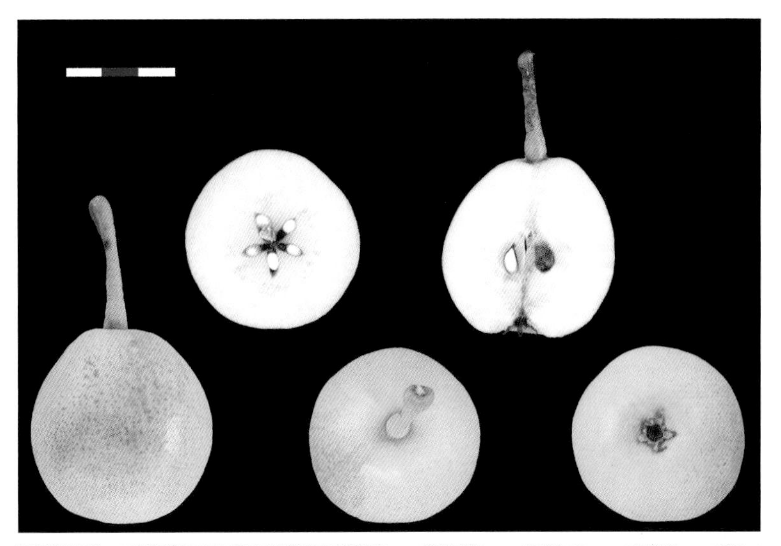

6. Seerkefu

Origin and Distribution　Seerkefu (2n=34), was originated in Xinjiang.

Main Characters　Tree: moderately vigorous or weak, medium productive. Leaf: 8.0cm × 4.7cm in size, elliptical. Initial leaf: green tinged with very light red. Flower: white bud, 6 to 9 flowers per cluster in average of 7.4; stamen number: 20 to 24, averaging 21.5; corolla diameter: 3.3cm. Fruit: matures in mid-September in Xingcheng, Liaoning Province, 64g per fruit, 4.9cm long, 4.9cm wide, sub-round or broadly obovate, greenish-yellow or yellow skin, some covered with light red on the side exposed to the sun; medium core, 5 locules, flesh white, tender, relatively fine, juicy or mid-juicy, sour-sweet or very sweet; TSS 13.43%, TA 0.13%; medium or above in quality.

7. 窝窝梨

来源及分布　2n=34，原产青海，在黄河流域有栽培。

主要性状　树势强，较丰产，始果年龄中等。叶片长 7.9cm，宽 6.0cm，阔卵圆形，初展叶绿色，微着红色，有茸毛。花蕾白色，边缘浅粉红色，每花序 6 ～ 8 朵花，平均 7.4 朵；雄蕊 21 ～ 30 枚，平均 24.8 枚；花冠直径 4.7cm。在辽宁兴城，果实 8 月下旬或 9 月上旬成熟，单果重 141g，纵径 5.9cm，横径 6.4cm，阔倒卵圆形，不规则；果皮绿黄色，果面凹凸不平；果心中大，5 心室；果肉白色，肉质中粗，松脆，汁液多或中多，味甜酸；含可溶性固形物 10.95%；品质中上等或中等。

7. Wowoli

Origin and Distribution Wowoli (2n=34), originated in Qinghai Province, is grown mainly along Yellow River basin.

Main Characters Tree: vigorous, relatively productive, medium precocity. Leaf: 7.9cm × 6.0cm in size, broadly ovate. Initial leaf: green tinged with light red, pubescent. Flower: white bud tinged with light pink on edge, 6 to 8 flowers per cluster in average of 7.4; stamen number: 21 to 30, averaging 24.8; corolla diameter: 4.7cm. Fruit: matures in late August or early September in Xingcheng, Liaoning Province, 141g per fruit, 5.9cm long, 6.4cm wide, broadly obovate, irregular, greenish-yellow skin, uneven, medium core, 5 locules, flesh white, mid-coarse, crisp tender, juicy or mid-juicy, sweet-sour; TSS 10.95%; medium or above in quality.

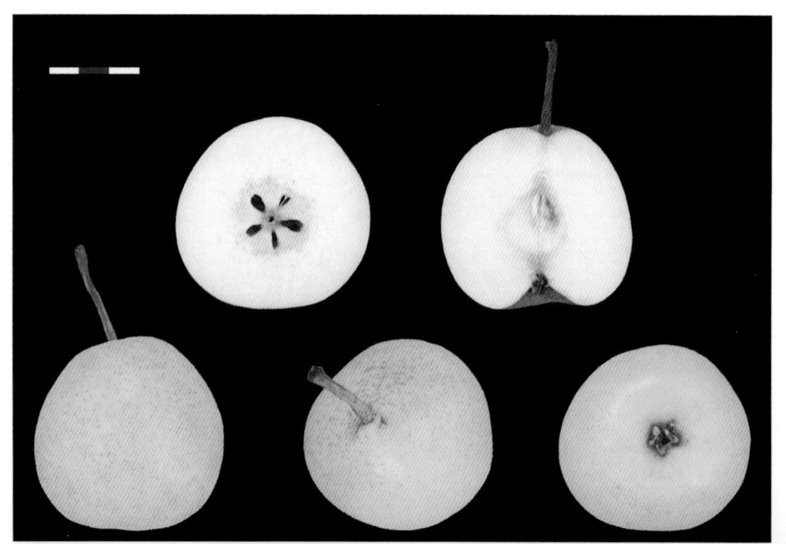

二、中国梨选育品种

Improved Pear Varieties in China

1. 八月红

来源及分布　原陕西省果树研究所和中国农业科学院果树研究所育成，母本为早巴梨，父本为早酥梨，1973 年杂交，1995 年通过审定。在陕西、辽宁等地有栽培。

主要性状　树势强，始果年龄早，丰产。叶片长 10.0cm，宽 5.5cm，椭圆形，初展叶绿色。花蕾白色，每花序 5 ~ 9 朵花，平均 6.5 朵；雄蕊 20 ~ 23 枚，平均 20.9 枚；花冠直径 3.7cm。在陕西关中地区，果实 8 月中旬成熟，单果重 233g，纵径 8.1cm，横径 7.2cm，卵圆形或圆锥形；果皮绿黄色或黄色，阳面有鲜红晕，外观较美；果心中大，5 心室；果肉白色，肉质细，脆，汁液多，味甜；含可溶性固形物 14.0%；品质上等；货架期 15d 左右。

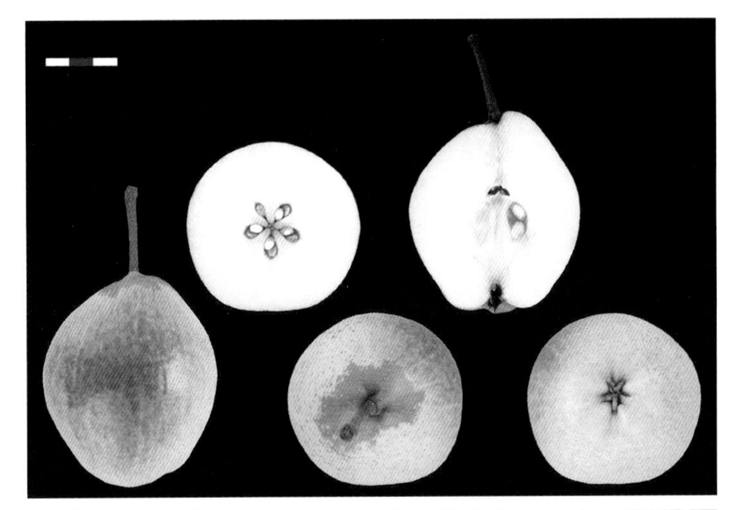

1. Bayuehong

Origin and Distribution　Bayuehong, was originated from a cross between Clapp Favorite (female parent) and Zaosu (male parent) which was made in 1973 by former Shaanxi Fruit Research Institute and Research Institute of Pomology, CAAS and released in 1995, is grown in Shaanxi, Liaoning, etc.

Main Characters　Tree: vigorous, productive, early precocity. Leaf: 10.0cm × 5.5cm in size, elliptical. Initial leaf: green. Flower: white bud, 5 to 9 flowers per cluster in average of 6.5; stamen number: 20 to 23, averaging 20.9; corolla diameter: 3.7cm. Fruit: matures in mid-August in the central Shaanxi plain, 233g per fruit, 8.1cm long, 7.2cm wide, ovate or conical, greenish-yellow or yellow skin covered with bright red on the side exposed to the sun, surface beautiful, medium core, 5 locules, flesh white, fine, crisp, juicy, sweet; TSS 14.0%; quality good; shelf life about 15 days.

2. 翠冠

来源及分布 浙江省农业科学院园艺研究所育成，母本为幸水，父本为杭青 × 新世纪，1979 年杂交，1999 年通过审定。浙江省主栽梨品种，在我国长江流域及以南地区有大面积栽培。

主要性状 树势强，丰产，始果年龄早。叶片长 13.8cm，宽 7.8cm，椭圆形，初展叶棕红色。花蕾白色，小花蕾边缘微显粉红色，每花序 6 ~ 9 朵花，平均 7.8 朵；雄蕊 20 ~ 29 枚，平均 22.3 枚；花冠直径 3.9cm。在浙江海宁，果实 7 月底至 8 月初成熟，单果重 277g，纵径 7.2cm，横径 8.2cm，圆形或扁圆形；果皮黄绿色，有果锈；果心较小，5 心室或 6 心室；果肉白色，肉质细，松脆，汁液极多，味甜，味浓；含可溶性固形物 13.20%，可滴定酸 0.11%；品质上等。

2. Cuiguan

Origin and Distribution Cuiguan, originated from a cross between Kousui (female parent)and Hangqing × Shinseiki (male parent) made in 1979 by the Horticultural Research Institute, Zhejiang Academy of Agricultural Sciences and released in 1999, is the main pear variety in Zhejiang, and cultivated widely along Yangtze River basin and its southern parts.

Main Characters Tree: vigorous, productive, early precocity. Leaf: 13.8cm × 7.8cm in size, elliptical. Initial leaf: brownish-red. Flower: white bud, small one tinged with light pink on edge, 6 to 9 flowers per cluster in average of 7.8; stamen number: 20 to 29, averaging 22.3; corolla diameter: 3.9cm. Fruit: matures in late July or early August in Haining, Zhejiang Province, 277g per fruit, 7.2cm, long, 8.2cm wide, round or oblate, yellowish-green skin, russet, relatively small core, 5 or 6 locules, flesh white, fine, tender and crisp, extremely juicy, sweet, rich flavor; TSS 13.20%, TA 0.11%; quality good.

3. 翠玉

来源及分布 浙江省农业科学院园艺研究所育成，母本为西子绿，父本为翠冠，1995 年杂交，2011 年通过审定。在浙江、福建、四川、江西、安徽等省有栽培。

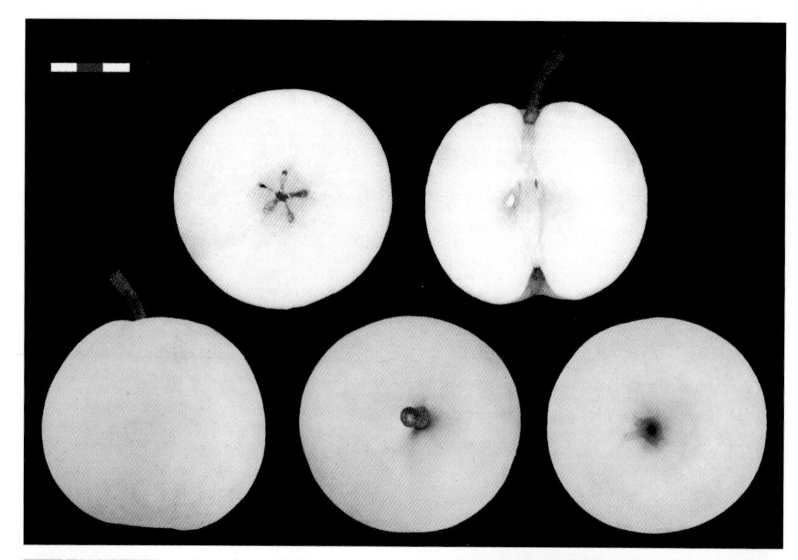

主要性状 树势中庸，丰产，始果年龄早。叶片长 13.0cm，宽 8.2cm，卵圆形，初展叶绿色。花蕾白色，每花序 4 ～ 8 朵花，平均 5.3 朵；雄蕊 16 ～ 36 枚，平均 24.5 枚。在浙江杭州，果实 7 月上、中旬成熟，单果重 304g，纵径 7.3cm，横径 8.5cm，扁圆形；果皮浅绿色或黄绿色，外观美；果心小，5 心室；果肉白色，肉质细，松脆或疏松，汁液多，味甜或淡甜；含可溶性固形物 11.0%，可滴定酸 0.16%；品质上等或中上等。

3. Cuiyu

Origin and Distribution Cuiyu, originated from a cross between Xizilü (female parent) and Cuiguan (male parent) made in 1995 by the Horticultural Research Institute, Zhejiang Academy of Agricultural Sciences and released in 2011, is grown in Zhejiang, Fujian, Sichuan, Jiangxi, Anhui, etc.

Main Characters Tree: vigorous, productive, early precocity. Leaf: 13.0cm × 8.2cm in size, ovate. Initial leaf: green. Flower: white bud, 4 to 8 flowers per cluster in average of 5.3; stamen number: 16 to 36, averaging 24.5. Fruit: matures in early or mid July in Hangzhou, Zhejiang Province, 304g per fruit, 7.3cm long, 8.5cm wide, oblate, light green or yellowish-green skin, surface beautiful, small core, 5 locules, flesh white, fine, crisp tender or tender, juicy, sweet or light sweet; TSS 11.0%, TA 0.16%; good or above medium in quality.

4. 初夏绿

来源及分布 浙江省农业科学院园艺研究所育成，母本为西子绿，父本为翠冠，1995 年杂交，2008 年通过品种认定。在浙江、江苏、江西、福建等省有栽培。

主要性状 树势强，丰产，始果年龄早。叶片长 13.4cm，宽 7.8cm，卵圆形，初展叶绿色。花蕾白色，每花序 5 ~ 8 朵花，平均 6.6 朵；雄蕊 19 ~ 35 枚，平均 28.2 枚。在浙江杭州，果实 7 月中旬成熟，较翠冠梨早 5d，货架期 10d 左右；单果重 278g，纵径 7.3cm，横径 8.3cm，扁圆形或近圆形；果皮浅绿色，无果锈，外观美；果心小或中大，5 心室；肉质细，松脆，汁液多，味甜；含可溶性固形物 10.0%；品质中上等或上等。

4. Chuxialü

Origin and Distribution Chuxialü, originated from a cross between Xizilü (female parent) and Cuiguan (male parent) by the Horticultural Research Institute, Zhejiang Academy of Agricultural Sciences made in 1995 and released in 2008, is grown in Zhejiang, Jiangsu, Jiangxi, Fujian, etc.

Main Characters Tree: vigorous, productive, early precocity. Leaf: 13.4cm × 7.8cm in size, ovate. Initial leaf: green. Flower: white bud, 5 to 8 flowers per cluster in average of 6.6; stamen number: 19 to 35, averaging 28.2. Fruit: matures in mid-July in Hangzhou, Zhejiang Province, 5 days earlier than Cui Guan, shelf life about 10 days, 278g per fruit, 7.3cm long, 8.3cm wide, oblate or sub-round, light green skin, free of russet, surface beautiful, small or medium core, 5 locules, flesh fine, crisp tender, juicy, sweet; TSS 10.0%; good or above medium in quality.

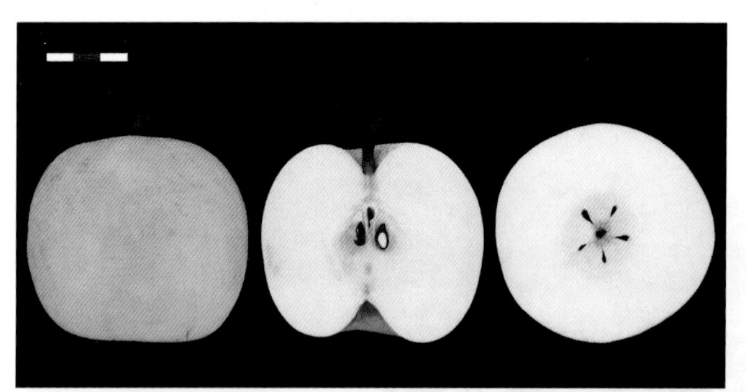

5. 大南果

来源及分布　鞍山市农林牧业局、鞍山副业总场、辽宁省果树科学研究所、沈阳农业大学选育，为南果梨芽变，1978 年在辽宁鞍山发现，1990 年通过审定。在辽宁、吉林、甘肃等地有栽培。

主要性状　树势强，丰产，始果年龄早或中等。叶片长 11.1cm，宽 6.9cm，卵圆形，初展叶绿色。花蕾白色，边缘粉红色，每花序 8 ～ 9 朵花，平均 8.7 朵；雄蕊 19 ～ 23 枚，平均 21.0 枚；花冠直径 3.3cm。在辽宁熊岳，果实 9 月上旬成熟，单果重 125g，扁圆形，部分果面有沟；果皮绿黄色，贮后转为黄色，阳面有淡红或鲜红晕；果心中大，5 心室；果肉淡黄白色，肉质细，紧密而脆，经 7 ～ 10d 后熟，果肉软溶，味酸甜，香气浓；含可溶性固形物 13.0%；品质极上。

5. Dananguo

Origin and Distribution　Dananguo, a bud mutation of Nanguoli, which was found in 1978 in Anshan, Liaoning Province by Agriculture, Forestry and Animal Husbandry Bureau of Anshan city, Anshan Steel Corp Subsidiary Factory, Liaoning Research Institute of Pomology, Shenyang Agricultural University, and approved in 1990, is grown in Liaoning, Jilin, Gansu, etc.

Main Characters　Tree: vigorous, productive, early or medium precocity. Leaf: 11.1cm × 6.9cm in size, ovate. Initial leaf: green. Flower: white bud tinged with pink on edge, 8 to 9 flowers per cluster in average of 8.7; stamen number: 19 to 23, averaging 21.0; corolla diameter: 3.3cm. Fruit: matures in early September in Xiongyue, Liaoning Province, 125g per fruit, oblate, some with grooves on surface, greenish-yellow skin, turning to yellow after storage, covered with light or bright red on the side exposed to the sun; medium core, 5 locules, flesh pale yellowish-white, fine, dense and crisp, turning to melting after 7 to 10 days of storage, sour-sweet, strongly aromatic; TSS 13.0%; extremely good in quality.

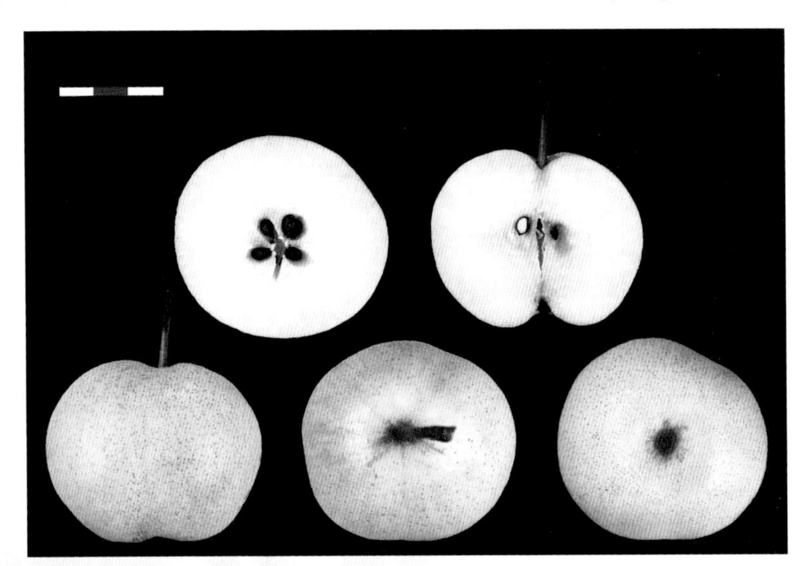

6. 大慈梨

来源及分布 吉林省农业科学院果树研究所育成，母本为大梨，父本为茌梨，1977 年杂交，1995 年通过审定。在吉林、辽宁、新疆等地有栽培。

主要性状 树势强，丰产，始果年龄早。叶片长 12.9cm，宽 6.9cm，卵圆形，初展叶红色，微显绿色。花蕾白色，边缘粉红色，每花序 4 ～ 5 朵花，平均 4.6 朵；雄蕊 20 枚；花冠直径 3.9cm。在吉林公主岭，果实 9 月下旬成熟，单果重 203g，纵径 7.4cm，横径 7.2cm，椭圆形、卵圆形或纺锤形，不整齐；果皮黄绿色，果点明显；果心中大，5 心室；果肉黄白色，肉质细，脆而稍紧密，汁液多，味甜酸；含可溶性固形物 13.0%；品质上等。果实耐贮藏。

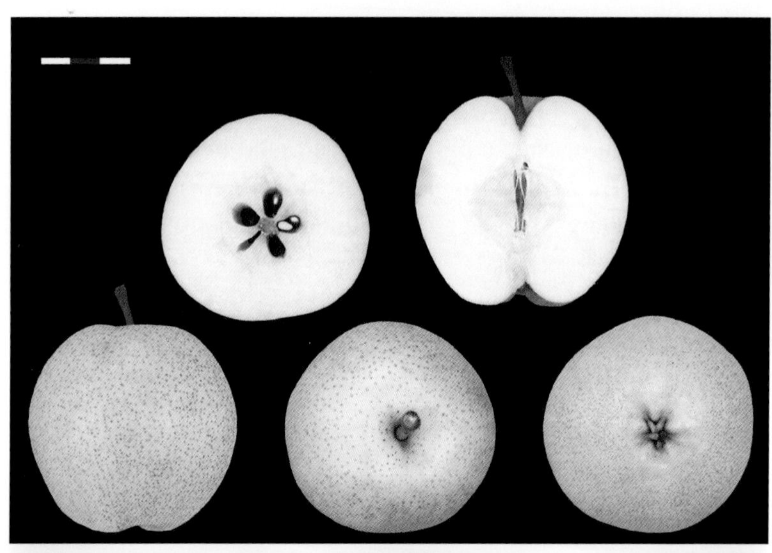

6. Dacili

Origin and Distribution Dacili, originated from a cross between Dali (female parent) and Chili (male parent) made in 1977 by Fruit Research Institute, Jilin Academy of Agricultural Sciences and released in 1995, is grown in Jilin, Liaoning and Xinjiang, etc.

Main Characters Tree: vigorous, productive, early precocity. Leaf: 12.9cm × 6.9cm in size, ovate. Initial leaf: red with less green. Flower: white bud tinged with pink on edge, 4 to 5 flowers per cluster in average of 4.6; stamens number 20; corolla diameter: 3.9cm. Fruit: matures in late September in Gongzhuling, Jilin Province, 203g per fruit, 7.4cm long, 7.2cm wide, elliptical, ovate, or spindle-shaped, irregular, yellowish-green skin with dots obvious, medium core, 5 locules, flesh yellowish-white, fine, crisp, somewhat dense, juicy, sweet-sour; TSS 13.0%; quality good; storage life long.

7. 大梨

来源及分布 吉林省农业科学院果树研究所育成，苹果梨实生，1956年采集苹果梨种子实生播种，1989年育成。在吉林、辽宁、河北、新疆等地有栽培。

主要性状 树势强，丰产，始果年龄中等。叶片长10.0cm，宽6.3cm，卵圆形，初展叶绿色，着暗红色。花蕾白色，边缘粉红色，每花序4～7朵花，平均5.4朵；雄蕊20～24枚，平均21.5枚；花冠直径4.0cm。在吉林公主岭，果实9月下旬成熟，单果重362g，纵径7.3cm，横径9.3cm，扁圆形，稍显不规则，果面凹凸不平；果皮绿黄色或黄色，向阳面有红晕；果心小，5心室；果肉雪白，肉质细，脆，汁液多，味甜酸，含可溶性固形物11.61%，品质上等。果实较耐贮藏，适宜鲜食及加工制罐。

7. Dali

Origin and Distribution Dali, originated from one of the seedlings of Pingguoli, seeds collected in 1956, and approved in 1989, is grown in Jilin, Liaoning, Hebei, Xinjiang, etc.

Main Characters Tree: vigorous, productive, medium precocity. Leaf: 10.0cm × 6.3cm in size, ovate. Initial leaf: green tinged with dark red. Flower: white bud tinged with pink on edge, 4 to 7 flowers per cluster in average of 5.4; stamen number: 20 to 24, averaging 21.5; corolla diameter: 4.0cm. Fruit: matures in late September in Gongzhuling, Jilin Province, 362g per fruit, 7.3cm long, 9.3cm wide, oblate, irregular, uneven, greenish-yellow or yellow skin covered with red on the side exposed to the sun, small core, 5 locules, flesh snowy-white, fine, crisp, juicy, sweet-sour; TSS 11.61%; quality good; storage life relatively long, suitable for fresh eating and canning.

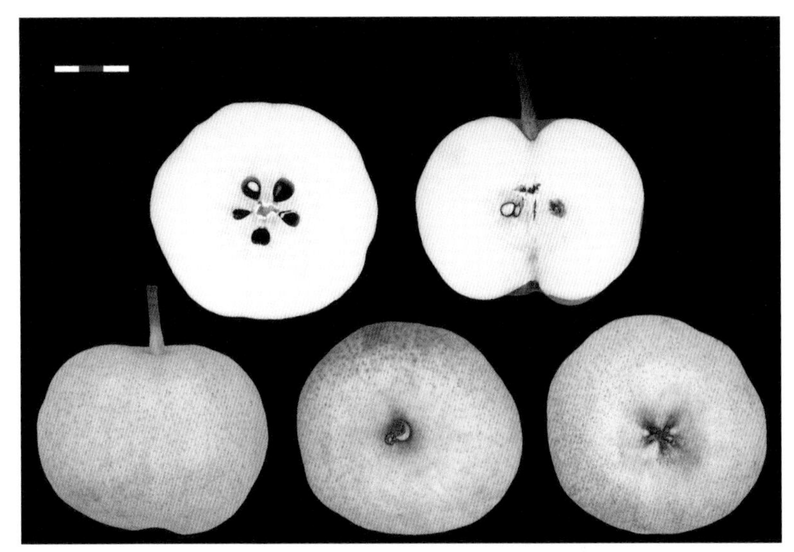

8. 冬蜜

来源及分布 黑龙江省农业科学院园艺分院育成，母本为龙香梨，父本为园月、库尔勒香梨、冬果梨3个品种混合花粉，1972年杂交，1999年通过审定。在黑龙江省东部和南部等地有栽培。

主要性状 树势中庸，丰产，始果年龄中等。叶片长11.8cm，宽8.1cm，阔椭圆形，初展叶绿色，着红色。花蕾白色，边缘淡粉红色，每花序6～9朵花，平均7.3朵；雄蕊25～30枚，平均27.7枚；花冠直径5.0cm。在黑龙江哈尔滨，果实9月底成熟，单果重140g，纵径7.0cm，横径6.2cm，圆形；果皮黄褐色；果心中大，5或4心室；果肉白色，肉质细，脆，稍紧密，汁液多，味酸甜；含可溶性固形物14.23%，可滴定酸0.31%；品质中上等。适宜冻藏。

8. Dongmi

Origin and Distribution Dongmi, was originated from a cross between Longxiangli (female parent) and mixed pollen, Yuanyue/Korla Pear/Dongguo, as male parent, which was made in 1972 and released in 1999 by Horticultural Branch, Heilongjiang Academy of Agricultural Sciences, is grown in the eastern and southern parts of Heilongjiang, etc.

Main Characters Tree: moderately vigorous, productive, medium precocity. Leaf: 11.8cm × 8.1cm in size, broadly elliptical. Initial leaf: green tinged with red. Flower: white bud tinged with light pink on edge, 6 to 9 flowers per cluster in average of 7.3; stamen number: 25 to 30, averaging 27.7; corolla diameter: 5.0cm. Fruit: matures in late September in Harbin, Heilongjiang Province, 140g per fruit, 7.0cm long, 6.2cm wide, round, brownish-yellow skin, medium core, 5 or 4 locules, flesh white, fine, crisp, somewhat dense, juicy, sour-sweet; TSS 14.23%, TA 0.31%; quality above medium; suitable for frozen pears.

9. 鄂梨 1 号

来源及分布 2n=34，湖北省农业科学院果树茶叶研究所育成，母本为伏梨，父本为金水酥，1982 年杂交，2002 年通过审定。

主要性状 树势中庸，丰产，始果年龄早。叶片长 10.6cm，宽 6.1cm，卵圆形，初展叶绿色。花蕾白色，每花序 6 ~ 8 朵花，平均 7.0 朵；雄蕊 19 ~ 25 枚，平均 21.3 枚。在湖北武汉，果实 7 月上旬成熟，单果重 220g，纵径 8.7cm，横径 8.5cm，近圆形；果皮绿色；果心小，5 心室；果肉白色，肉质细，脆，汁液多，味甜；含可溶性固形物 10.6%，可滴定酸 0.22%；品质中上等。果实不耐贮藏。

9. Eli No.1

Origin and Distribution Eli No.1 (2n=34), originated from a cross between Fuli (female parent) and Jinshuisu (male parent) made in 1982 by Institute of Fruit and Tea, Hubei Academy of Agricultural Sciences, was approved in 2002.

Main Characters Tree: moderately vigorous, productive, early precocity. Leaf: 10.6cm × 6.1cm in size, ovate. Initial leaf: green. Flower: white bud, 6 to 8 flowers per cluster in average of 7.0; stamen number: 19 to 25, averaging 21.3. Fruit: matures in early July in Wuhan, Hubei Province, 220g per fruit, 8.7cm long, 8.5cm wide, sub-round, green skin, small core, 5 locules, flesh white, fine, crisp, juicy, sweet; TSS 10.6%, TA 0.22%; quality above medium; storage life short.

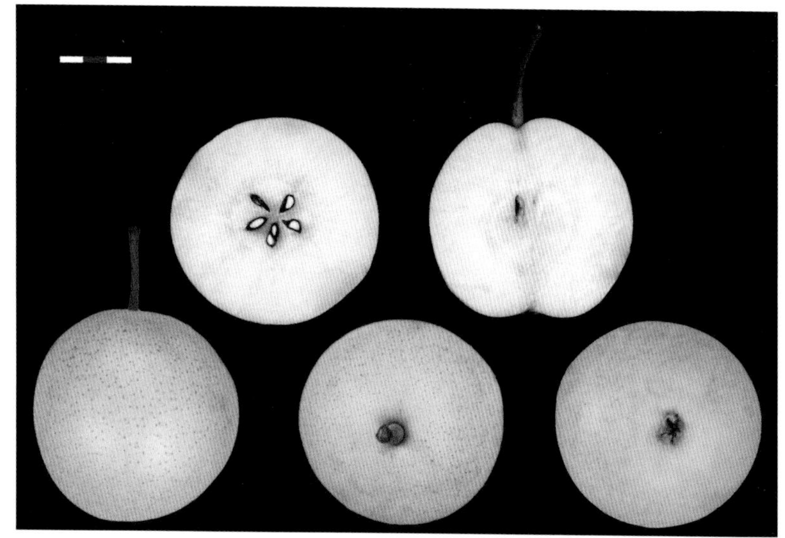

10. 鄂梨2号

来源及分布 2n=34，湖北省农业科学院果树茶叶研究所育成，母本为中香，父本为伏梨 × 启发，1982年杂交，2002年通过审定。

主要性状 树势强，丰产，始果年龄早。叶片长11.5cm，宽6.6cm，狭椭圆形，初展叶橙红色。花蕾白色，边缘粉红色，每花序4 ~ 7朵花，平均6.1朵；雄蕊20 ~ 25枚，平均22.6枚。果实7月中、下旬成熟，单果重194g，纵径8.4cm，横径7.2cm，倒卵圆形；果皮黄绿色；果心小，5心室；果肉白色，肉质细，疏松，汁液特多，味甜；含可溶性固形物12.1%，可滴定酸0.22%，品质上等。果实不耐贮藏。

10. Eli No.2

Origin and Distribution Eli No.2 (2n=34), originated from a cross between Zhongxiang (female parent) and Fuli × Qifa (male parent) made in 1982 by Institute of Fruit and Tea, Hubei Academy of Agricultural Sciences, was approved in 2002.

Main Characters Tree: vigorous, productive, early precocity. Leaf: 11.5cm × 6.6cm in size, narrowly elliptical. Initial leaf: orange red. Flower: white bud tinged with pink on edge, 4 to 7 flowers per cluster in average of 6.1; stamen number: 20 to 25, averaging 22.6. Fruit: matures in mid or late July, 194g per fruit, 8.4cm long, 7.2cm wide, obovate, yellowish-green skin, small core, 5 locules, flesh white, fine, tender, extremely juicy, sweet; TSS 12.1%, TA 0.22%; quality good; storage life short.

11. 甘梨早 8

来源及分布 甘肃省农业科学院林果花卉研究所育成，母本为四百目，父本为早酥，1981 年杂交，2008 年通过审定。在甘肃省天水、平凉、兰州、白银以及河西等地有栽培。

主要性状 树势强，丰产，始果年龄早。叶片长 13.0cm，宽 9.5cm，阔卵圆形，初展叶绿色。花蕾白色，小花蕾边缘浅粉红色，每花序 5 ~ 9 朵花，平均 6.5 朵；雄蕊 20 ~ 24 枚，平均 21.4 枚；花冠直径 3.8cm。在甘肃兰州，果实 8 月初成熟，单果重 256g，纵径 8.3cm，横径 7.5cm，卵圆形；果皮黄绿色或黄色；果心小，5 心室；果肉白色，肉质细，松脆，汁液多，味甜；含可溶性固形物 12.6%；品质上等或中上等。

11. Ganlizao 8

Origin and Distribution Ganlizao 8, originated from a cross between Sibaimu (female parent) and Zaosu (male parent) made by Institute of Fruit and Floriculture, Gansu Academy of Agricultural Sciences in 1981 and released in 2008, is grown in Tianshui, Pingliang, Lanzhou, Baiyin, and Hexi, etc.

Main Characters Tree: vigorous, productive, early precocity. Leaf: 13.0cm × 9.5cm in size, broadly ovate. Initial leaf: green. Flower: white bud, small one tinged with light pink on edge, 5 to 9 flowers per cluster in average of 6.5; stamen number: 20 to 24, averaging 21.4; corolla diameter: 3.8cm. Fruit: matures in early August in Lanzhou, Gansu Province, 256g per fruit, 8.3cm long, 7.5cm wide, ovate, yellowish-green or yellow skin, small core, 5 locules, flesh white, fine, crisp tender, juicy, sweet; TSS 12.6%; good or above medium in quality.

12. 寒红梨

来源及分布 吉林省农业科学院果树研究所育成，母本为南果梨，父本为晋酥梨，2003 年通过审定。在辽宁、吉林、黑龙江、内蒙古等地有栽培。

主要性状 树势中庸，丰产，始果年龄中等。叶片长 10.5cm，宽 6.3cm，卵圆形，初展叶绿色，微着红色。花蕾白色，边缘淡粉红色，每花序 6 ~ 7 朵花，平均 6.3 朵；雄蕊 20 ~ 29 枚，平均 24.0 枚；花冠直径 4.6cm。在吉林公主岭，果实 9 月下旬成熟，单果重 200g，纵径 8.0cm，横径 7.9cm，圆形或阔纺锤形；果皮绿黄色或黄色，阳面有红晕，外观美；果心中大，心室数 4 或 5；果肉淡黄白色，肉质细，脆，汁液多，味甜酸，微有香气；含可溶性固形物 14.0%；品质上等。

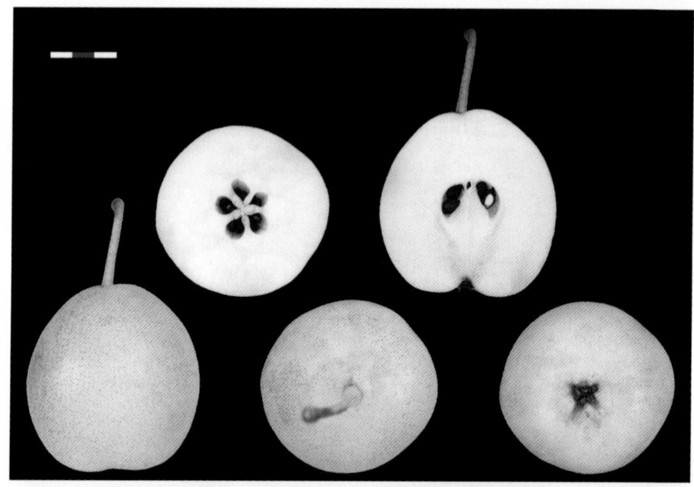

12. Hanhongli

Origin and Distribution Hanhongli, originated from a cross between Nanguoli (female parent) and Jinsuli (male parent) by Fruit Research Institute, Jilin Academy of A gricultural Sciences and released in 2003, is grown in Liaoning, Jilin, Heilongjiang, Inner Mongolia, etc.

Main Characters Tree: moderately vigorous, productive, medium precocity. Leaf: 10.5cm × 6.3cm in size, ovate. Initial leaf: green tinged with light red. Flower: white bud tinged with light pink on edge, 6 to 7 flowers per cluster in average of 6.3; stamen number: 20 to 29, averaging 24.0; corolla diameter: 4.6cm. Fruit: matures in late September in Gongzhuling, Jilin Province, 200g per fruit, 8.0cm long, 7.9cm wide, round or broadly spindle-shaped, greenish-yellow or yellow skin covered with red on the side exposed to the sun, beautiful, medium core, 4 or 5 locules, flesh pale yellowish-white, fine, crisp, juicy, sweet-sour, lightly aromatic; TSS 14.0%; quality good.

13. 寒香梨

来源及分布 吉林省农业科学院果树研究所育成，母本为延边大香水，父本为苹香梨，1972 年杂交，2002 年通过审定。在辽宁、吉林、黑龙江、内蒙古等地有栽培。

主要性状 树势强，丰产，始果年龄中等。叶片长 12.7cm，宽 6.9cm，椭圆形，初展叶绿色，微着红色。花蕾白色，每花序 7 ~ 8 朵花，平均 7.2 朵；雄蕊 20 ~ 24 枚，平均 20.7 枚；花冠直径 4.1cm。在吉林中部地区，果实 9 月下旬成熟，单果重 151g，纵径 6.6cm，横径 6.5cm，近圆形；果皮黄绿色或黄色，部分果实阳面有红晕；果心中大偏小，5 心室；果肉白色，采收时肉质硬，经 10d 左右后熟，肉质变软，细，汁液多，酸甜味浓，有香气；含可溶性固形物 14.38%，可滴定酸 0.32%；品质上等。

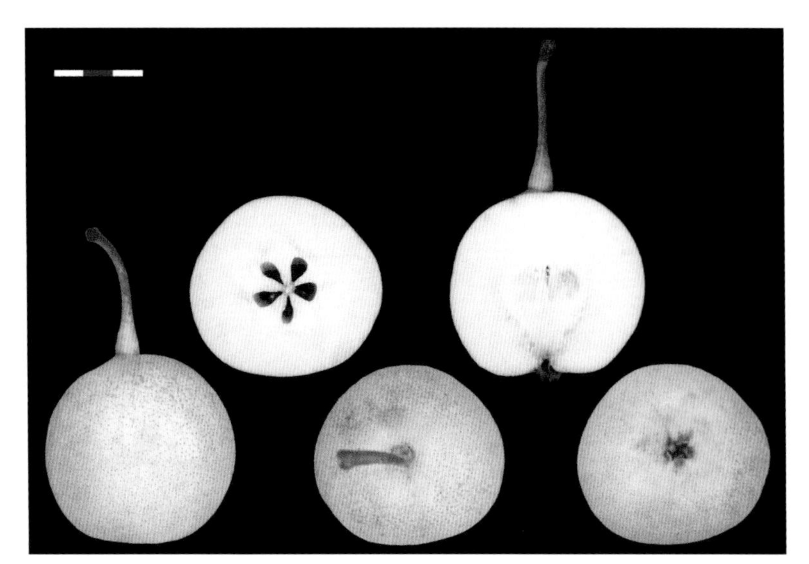

13. Hanxiangli

Origin and Distribution Hanxiangli, originated from a cross between Yanbian Daxiangshui (female parent) and Pingxiangli (male parent) made by Fruit Research Institute, Jilin Academy of Agricultural Sciences in 1972 and released in 2002, is grown in Liaoning, Jilin, Heilongjiang, Inner Mongolia,etc.

Main Characters Tree: vigorous, productive, medium precocity. Leaf: 12.7cm × 6.9cm in size, elliptical. Initial leaf: green tinged with light red. Flower: white bud, 7 to 8 flowers per cluster in average of 7.2; stamen number: 20 to 24, averaging 20.7; corolla diameter: 4.1cm. Fruit: matures in late September in central Jilin, 151g per fruit, 6.6cm long, 6.5cm wide, sub-round, yellowish-green or yellow skin, some covered with red on the side exposed to the sun, medium core, tending to small, 5 locules, flesh white, firm when harvested, becoming soft after 10 days of storage, fine, juicy, strongly sour-sweet, aromatic; TSS 14.38%, TA 0.32%; quality good.

14. 红香酥

来源及分布 中国农业科学院郑州果树研究所育成，母本为库尔勒香梨，父本为郑州鹅梨，1980 年杂交，1997 年通过审定。在黄河流域、云贵高原、华北平原等地有栽培。

主要性状 树势中庸，丰产，始果年龄早。叶片长 10.4cm，宽 7.0cm，卵圆形，初展叶红色。花蕾白色，边缘浅粉红色，每花序 6 ~ 7 朵花，平均 6.4 朵；雄蕊 20 ~ 25 枚，平均 22.5 枚；花冠直径 3.5cm。在河南郑州，果实 9 月中、下旬成熟，单果重 220g，纺锤形或长卵圆形；果皮绿黄色，阳面有红晕；果心中大，5 或 4 心室；果肉淡黄白色，肉质较细或中粗，松脆，汁液多，味甜；含可溶性固形物 13.5%；品质上等。果实较耐贮藏。

14. Hongxiangsu

Origin and Distribution Hongxiangsu, originated from a cross between Korla Pear (female parent) and Zhengzhou Eli (male parent) made in 1980 by Zhengzhou Fruit Research Institute, Chinese Academy of Agricultural Sciences and released in 1997, is grown along Yellow River basin, Yunnan-Guizhou Plateau, North China Plain, etc.

Main Characters Tree: moderately vigorous, productive, early precocity. Leaf: 10.4cm × 7.0cm in size, ovate. Initial leaf: red. Flower: white bud tinged with light pink on edge, 6 to 7 flowers per cluster in average of 6.4; stamen number: 20 to 25, averaging 22.5; corolla diameter: 3.5cm. Fruit: matures in mid or late September in Zhengzhou, Henan Province, 220g per fruit, spindle-shaped or elongated ovate, greenish-yellow skin covered with red on the side exposed to the sun, medium core, 5 or 4 locules, flesh pale yellowish-white, relatively fine or mid-coarse, crisp tender, juicy, sweet; TSS 13.5%; quality good; storage life relatively long.

15. 红秀 2 号

来源及分布 新疆生产建设兵团农七师果树研究所育成，母本为大香水，父本为苹果梨，1975 年杂交，1989 年通过鉴定。在北疆有栽培。

主要性状 树势中庸，丰产，始果年龄早。叶片长 12.0cm，宽 7.3cm，卵圆形，初展叶绿色，着红色。花蕾白色，或边缘着浅粉红色，每花序 7 ~ 10 朵花，平均 8.2 朵；雄蕊 20 ~ 23 枚，平均 20.9 枚；花冠直径 3.5cm。在新疆奎屯，果实 9 月中旬成熟，单果重 156g，纵径 6.1cm，横径 7.1cm，扁圆形，不规则；果皮绿黄色，阳面有红晕；果心中大，5 心室；果肉白色，采收时肉质脆，经 7d 左右变软，中粗，汁液多，味酸甜；含可溶性固形物 11.74%；品质中上等。

15. Hongxiu No.2

Origin and Distribution Hongxiu No.2, was originated from a cross between Daxiangshui (female parent) and Pingguoli (male parent) made in 1975 by Fruit Research Institute, Agronomy Seventh Division, Xinjiang Production and Construction Corps, and released in 1989, is grown in Northern Xinjiang.

Main Characters Tree: moderately vigorous, productive, early precocity. Leaf: 12.0cm × 7.3cm in size, ovate. Initial leaf: green tinged with red. Flower: white bud or tinged with light pink on edge, 7 to 10 flowers per cluster in average of 8.2; stamen number: 20 to 23, averaging 20.9; corolla diameter: 3.5cm. Fruit: matures in mid-September in Kuitun, Xinjiang, 156g per fruit, 6.1cm long, 7.1cm wide, oblate, irregular, greenish-yellow skin, medium core, 5 locules, flesh white, crisp when harvested, turning to soft after 7 days of storage, mid-coarse, juicy, sour-sweet; TSS 11.74%; quality above medium.

16. 华梨 1 号

来源及分布 2n=34，华中农业大学育成，母本为湘南，父本为江岛，1978 年杂交，1997 年通过审定。在湖北、湖南等地有栽培。

主要性状 树势强，丰产，始果年龄早。叶片长 9.9cm，宽 6.0cm，长卵圆形，初展叶橘红色。花蕾白色，边缘粉红色，每花序 2 ~ 9 朵花，平均 5.4 朵；雄蕊 19 ~ 24 枚，平均 21.6 枚。在湖北武汉，果实 9 月上旬成熟，单果重 292g，纵径 7.1cm，横径 8.1cm，广卵圆形；果皮浅褐色；果心中大，5 心室；果肉白色，肉质细，松脆，汁液多，味酸甜；含可溶性固形物 11.8%，可滴定酸 0.11%；品质中上等。

16. Huali No.1

Origin and Distribution Huali No.1 (2n=34), originated from a cross between Shounan (female parent) and Enoshima (male parent) made in 1978 by Huazhong Agricultural University and released in 1997, is grown in Hubei, Hunan, etc.

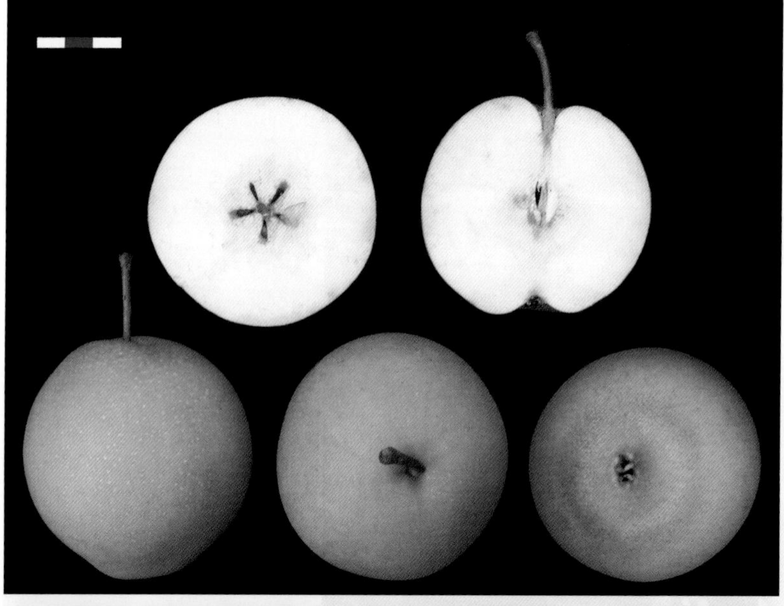

Main Characters Tree: vigorous, productive, early precocity. Leaf: 9.9cm × 6.0cm in size, elongated ovate. Initial leaf: orange red. Flower: white bud tinged with pink on edge, 2 to 9 flowers per cluster in average of 5.4; stamen number: 19 to 24, averaging 21.6. Fruit: matures in early September in Wuhan, Hubei Province, 292g per fruit, 7.1cm long, 8.1cm wide, broadly ovate, light brown skin, medium core, 5 locules, flesh white, fine, crisp tender, juicy, sour-sweet; TSS 11.8%, TA 0.11%; quality above medium.

17. 华金

来源及分布 中国农业科学院果树研究所育成，母本为早酥，父本为早白，2003 年获植物新品种权。在辽宁、河北等地有栽培。

主要性状 树势较强，丰产，始果年龄早。叶片长 12.4cm，宽 7.2cm，椭圆形，初展叶绿色，微着红色。花蕾白色，每花序 5 ~ 8 朵花，平均 6.3 朵；雄蕊 18 ~ 21 枚，平均 19.6 枚；花冠直径 4.0cm。在辽宁兴城，果实 8 月上、中旬成熟，单果重 305g，长圆形或长卵圆形；果心中大偏小，5 心室；果肉淡黄白色，肉质松脆，较细，汁液多，味甜；含可溶性固形物 11.0%；品质上等。

17. Huajin

Origin and Distribution Huajin, originated from a cross between Zaosu (female parent) and Zaobai (male parent) by Research Institute of Pomology, Chinese Academy of Agricultural Sciences and released in 2003, is grown in Liaoning, Hebei, etc.

Main Characters Tree: relatively vigorous, productive, early precocity. Leaf: 12.4cm × 7.2cm in size, elliptical. Initial leaf: green tinged with light red. Flower: white bud, 5 to 8 flowers per cluster in average of 6.3; stamen number: 18 to 21, averaging 19.6; corolla diameter: 4.0cm. Fruit: matures in early or mid August in Xingcheng, Liaoning Province, 305g per fruit, long-round or long-ovate, medium core, tending to small, 5 locules, flesh pale yellowish-white, crisp tender, relatively fine, juicy, sweet; TSS 11.0%; quality good.

18. 华酥

来源及分布 中国农业科学院果树研究所育成，母本为早酥，父本为八云，1977 年杂交，1999 年通过审定。在辽宁、河北、江苏、四川等地有栽培。

主要性状 树势中庸，丰产，始果年龄早。叶片长 11.6cm，宽 7.0cm，卵圆形，初展叶绿色，微着红色。花蕾白色，小花蕾边缘浅粉红色，每花序 5 ～ 7 朵花，平均 6.2 朵；雄蕊 24 ～ 29 枚，平均 26.1 枚；花冠直径 3.8cm。在辽宁兴城，果实 8 月上旬成熟，单果重 242g，纵径 7.2cm，横径 7.7cm，圆形；果皮绿色或黄绿色；果心中大偏小，5 心室；果肉淡黄白色，肉质细，松脆，汁液多，味酸甜；含可溶性固形物 10.5%；品质上等。

18. Huasu

Origin and Distribution Huasu, originated from a cross between Zaosu (female parent) and Yakumo (male parent) made by Research Institute of Pomology, Chinese Academy of Agricultural Sciences in 1977 and released in 1999, is grown in Liaoning, Hebei, Jiangsu, Sichuan, etc.

Main Characters Tree: moderately vigorous, productive, early precocity. Leaf: 11.6cm × 7.0cm in size, ovate. Initial leaf: green tinged with light red. Flower: white bud, small one tinged with light pink on edge, 5 to 7 flowers per cluster in average of 6.2; stamen number: 24 to 29, averaging 26.1; corolla diameter: 3.8cm. Fruit: matures in early August in Xingcheng, Liaoning Province, 242g per fruit, 7.2cm long, 7.7cm wide, round, green or yellowish-green skin, medium core, tending to small, 5 locules, flesh pale yellowish-white, fine, crisp tender, juicy, sour-sweet; TSS 10.5%; quality good.

19. 黄花

来源及分布 2n=34，原浙江农业大学育成，母本为黄蜜，父本为三花，1962 年杂交，1974 年通过鉴定。在浙江省有大面积栽培，在福建、湖北、江苏等地亦有栽培。

主要性状 树势强，丰产，始果年龄早。叶片长 10.2cm，宽 6.3cm，卵圆形，初展叶红色，微显绿色。花蕾白色，边缘浅粉红色，每花序 4 ～ 8 朵花，平均 6.1 朵；雄蕊 18 ～ 23 枚，平均 20.6 枚；花冠直径 3.7cm。在湖北武汉，果实 8 月中旬成熟，单果重 216g，纵径 6.2cm，横径 7.1cm，圆锥形；果皮黄褐色；果心中大，5 心室；果肉白色，肉质细，脆，汁液多，味甜；含可溶性固形物 11.7%，可滴定酸 0.19%；品质中上等。果实较耐贮藏。

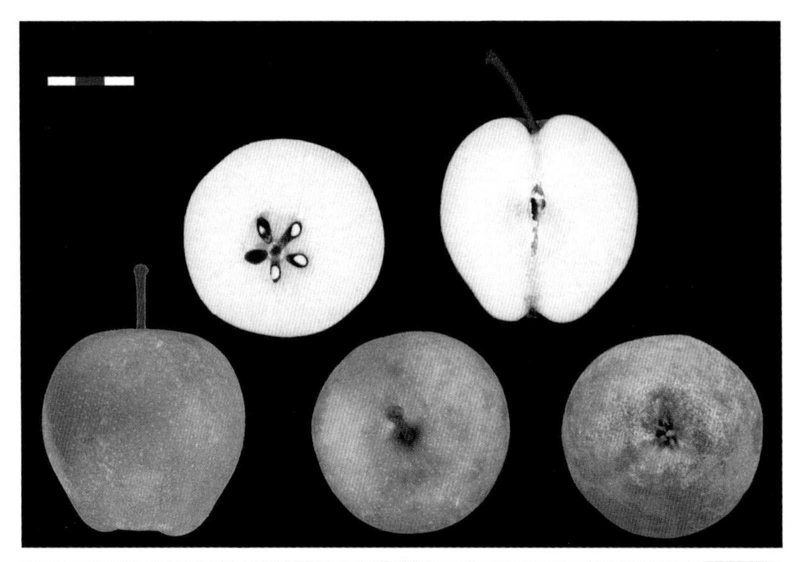

19. Huanghua

Origin and Distribution Huanghua (2n=34), originated from a cross between Imamuranatsu (female parent) and Sanhua (male parent) made by former Zhejiang Agricultural University in 1962 and released in 1974, is grown widely in Zhejiang Province, also grown in Fujian, Hubei and Jiangsu, etc.

Main Characters Tree: vigorous, productive, early precocity. Leaf: 10.2cm × 6.3cm in size, ovate. Initial leaf: red with less green. Flower: white bud tinged with light pink on edge, 4 to 8 flowers per cluster in average of 6.1; stamen number: 18 to 23, averaging 20.6; corolla diameter: 3.7cm. Fruit: matures in mid August in Wuhan, Hubei Province, 216g per fruit, 6.2cm long, 7.1cm wide, conical, russet skin, medium core, 5 locules, flesh white, fine, crisp, juicy, sweet; TSS 11.7%, TA 0.19%; quality above medium; storage life relatively long.

20. 黄冠

来源及分布 河北省农林科学院石家庄果树研究所育成，母本为雪花梨，父本为新世纪，1977 年杂交，1997 年通过审定。在我国华北、长江流域及其以南、西部地区等有大面积栽培。

主要性状 树势强，丰产，始果年龄早。叶片长 12.9cm，宽 7.6cm，卵圆形，初展叶红色。花蕾白色，每花序 6 ～ 8 朵花，平均 6.9 朵；雄蕊 18 ～ 29 枚，平均 22.2 枚；花冠直径 3.6cm。在河北晋州，果实 8 月中旬成熟，单果重 355g，纵径 8.3cm，横径 8.6cm，椭圆形，外观较美；果皮黄色或绿黄色；果心小，5 心室；果肉白色，肉质细，松脆，汁液多，味甜；含可溶性固形物 12.18%，可滴定酸 0.16%；品质上等。

20. Huangguan

Origin and Distribution Huangguan, originated from a cross between Xuehuali (female parent) and Shinseiki (male parent) made by Shijiazhuang Fruit Research Institute, Hebei Academy of Agro-Forestry Sciences in 1977 and released in 1997, is grown widely in North China, along and south to the Yangtze River basin, and West China, etc.

Main Characters Tree: vigorous, productive, early precocity. Leaf: 12.9cm × 7.6cm in size, ovate. Initial leaf: red. Flower: white bud, 6 to 8 flowers per cluster in average of 6.9; stamen number: 18 to 29, averaging 22.2; corolla diameter: 3.6cm. Fruit: matures in mid August in Jinzhou, Hebei Province, 355g per fruit, 8.3cm long, 8.6cm wide, elliptical, surface beautiful, yellow or greenish-yellow skin, small core, 5 locules, flesh white, fine, crisp tender, juicy, sweet; TSS 12.18%, TA 0.16%; quality good.

21. 锦丰

来源及分布 中国农业科学院果树研究所育成，母本为苹果梨，父本为茌梨，1956 年杂交，1969 年定名。在我国东北西部、华北北部及西北等地有栽培。

主要性状 树势强，产量较高，始果年龄晚。叶片长 14.5cm，宽 7.8cm，卵圆形，初展叶绿色，着红色。花蕾白色，边缘粉红色，每花序 6 ~ 7 朵花，平均 6.6 朵；雄蕊 20 ~ 21 枚，平均 20.2 枚；花冠直径 4.4cm。在辽宁兴城，果实 10 月上旬成熟，单果重 282g，纵径 7.6cm，横径 8.0cm，近圆形；果皮绿黄色，贮藏后转黄色，果点大而明显；果心小，5 心室；果肉白色，肉质细，松脆，汁液多，酸甜味浓；含可溶性固形物 13.53%；品质极上。果实耐贮藏。

21. Jinfeng

Origin and Distribution Jinfeng, originated from a cross between Pingguoli (female parent) and Chili (male parent) made by Research Institute of Pomology, Chinese Academy of Agricultural Sciences in 1956 and released in 1969, is grown in western part of Northeast China, northern part of North China, and Northwest China.

Main Characters Tree: vigorous, relatively productive, late precocity. Leaf: 14.5cm × 7.8cm in size, ovate. Initial leaf: green tinged with red. Flower: white bud tinged with pink on edge, 6 to 7 flowers per cluster in average of 6.6; stamen number: 20 to 21, averaging 20.2; corolla diameter: 4.4cm. Fruit: matures in early October in Xingcheng, Liaoning Province, 282g per fruit, 7.6cm long, 8.0cm wide, sub-round, greenish-yellow skin with dots large and obvious, turning to yellow after storage; small core, 5 locules, flesh white, fine, crisp tender, juicy, strongly sour-sweet; TSS 13.53%; extremely good in quality; storage life long.

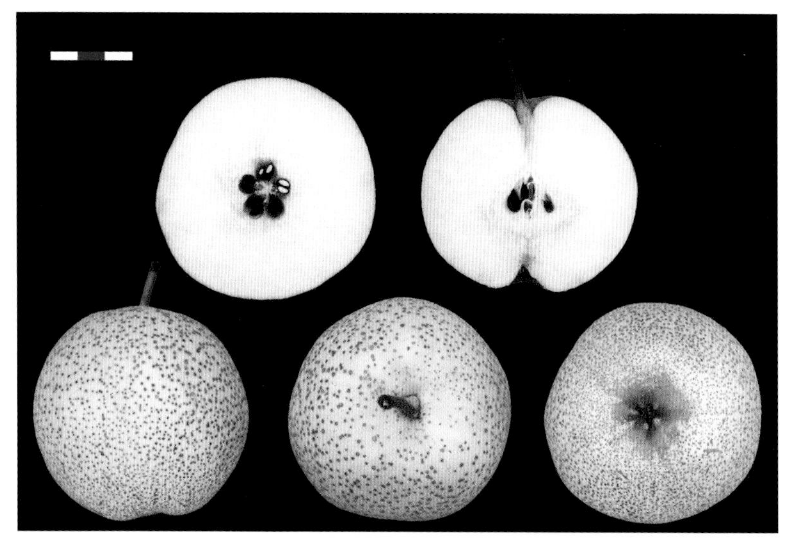

22. 锦香

来源及分布 中国农业科学院果树研究所育成，母本为南果梨，父本为巴梨，1956年杂交，1989年通过专家鉴定，2003年获植物新品种权。在辽宁、河北等地有栽培。

主要性状 树势中庸，产量中等，始果年龄早。叶片长6.4cm，宽4.5cm，卵圆形，初展叶绿色，微着浅红色。花蕾白色，每花序6～7朵花，平均6.5朵；雄蕊20～27枚，平均21.5枚；花冠直径3.8cm。在辽宁兴城，果实9月中旬成熟，单果重128g，纵径6.8cm，横径6.2cm，纺锤形；果皮绿黄色或黄色，阳面有红晕；果心中大，5心室；果肉淡黄白色，采收时肉质坚硬，不能食用，经后熟变软溶，细，汁液多，味甜酸，具浓香；含可溶性固形物13.73%；品质上等。适宜加工制罐。

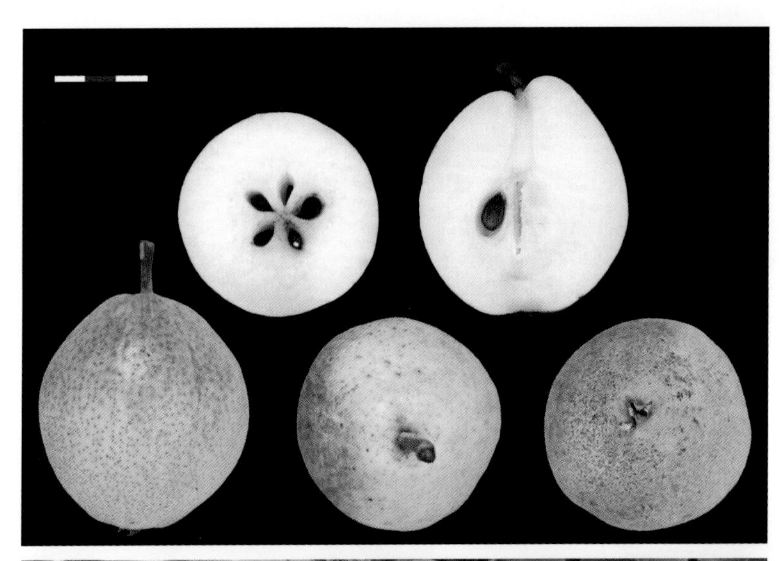

22. Jinxiang

Origin and Distribution Jinxiang, originated from a cross between Nanguoli (female parent) and Bartlett (male parent) made by Research Institute of Pomology, Chinese Academy of Agricultural Sciences in 1956, released in 1989, and granted new plant variety right in 2003, is grown in Liaoning, Hebei, etc.

Main Characters Tree: moderately vigorous, medium productive, early precocity. Leaf: 6.4cm × 4.5cm in size, ovate. Initial leaf: green slightly tinged with light red. Flower: white bud, 6 to 7 flowers per cluster in average of 6.5; stamen number: 20 to 27, averaging 21.5; corolla diameter: 3.8cm. Fruit: matures in mid-September in Xingcheng, Liaoning Province, 128g per fruit, 6.8cm long, 6.2cm wide, spindle-shaped, greenish-yellow or yellow skin covered with red on the side exposed to the sun, medium core, 5 locules, flesh pale yellowish-white, firm when harvested, inedible, turning to melting after storage, fine, juicy, sweet-sour, strongly aromatic; TSS 13.73%; quality good; suitable for canning.

23. 晋蜜梨

来源及分布 山西省农业科学院果树研究所育成，母本为砀山酥梨，父本为猪嘴梨，1972 年杂交，1985 年通过鉴定。在山西、内蒙古等地有栽培。

主要性状 树势中庸，较丰产，始果年龄中等。叶片长 8 ～ 12cm，宽 7 ～ 9cm，阔卵圆形，初展叶绿色，微着红色，有茸毛。花蕾白色，边缘淡粉红色，每序 6 ～ 8 朵花，平均 6.9 朵；雄蕊 20 ～ 26 枚，平均 21.9 枚；花冠直径 4.9cm。在晋中地区，果实 9 月底或 10 月上旬成熟，单果重 230g，卵圆形或椭圆形；果皮绿黄色或黄色，果点明显；果心小，3 ～ 5 心室；果肉白色，肉质细，松脆，汁液多，味甘甜；含可溶性固形物 16.0%；品质上等。果实耐贮藏。

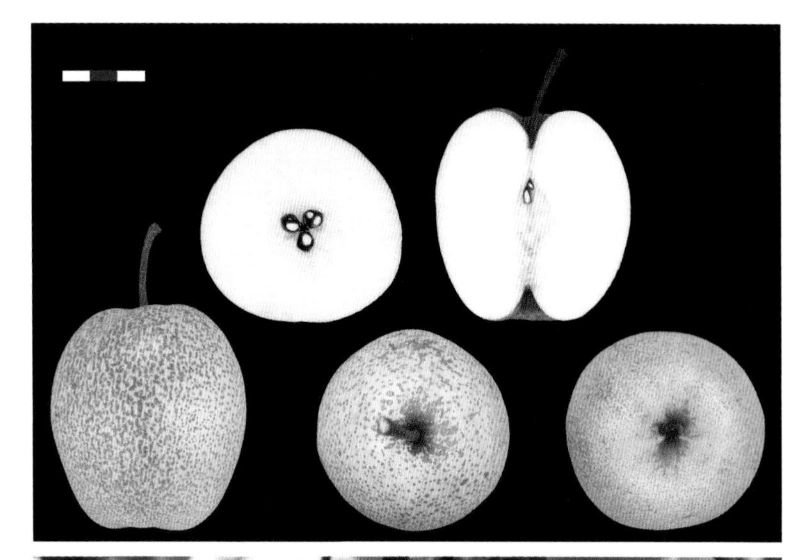

23. Jinmili

Origin and Distribution Jinmili, originated from a cross between Dangshan Suli (female parent) and Zhuzuili (male parent) made by Fruit Research Institute, Shanxi Academy of Agricultural Sciences in 1972 and released in 1985, is grown in Shanxi, Inner Mongolia, etc.

Main Characters Tree: moderately vigorous, relatively productive, medium precocity. Leaf: (8-12)cm × (7-9)cm in size, broadly ovate. Initial leaf: green tinged with light red, pubescent. Flower: white bud tinged with light pink on edge, 6 to 8 flowers per cluster in average of 6.9; stamen number: 20 to 26, averaging 21.9; corolla diameter: 4.9cm. Fruit: matures in late September or early October in central Shanxi Province, 230g per fruit, ovate or elliptical, greenish-yellow or yellow skin with dots obvious, small core, 3 to 5 locule, flesh white, fine, crisp tender, juicy, very sweet; TSS 16.0%; quality good; storage life long.

24. 晋酥梨

来源及分布 山西省农业科学院果树研究所育成，母本为鸭梨，父本为金梨，1957年杂交，1972年命名。在山西、陕西、云南等地有栽培。

主要性状 树势中庸，丰产，始果年龄早或中等。叶片长10.4cm，宽7.4cm，卵圆形或阔卵圆形，初展叶红色，微显绿色。花蕾白色，边缘粉红色，花瓣边缘亦粉红色，每花序5～7朵花，平均6.2朵；雄蕊20～26枚，平均21.5枚；花冠直径3.7cm。在山西晋中地区，果实9月中、下旬成熟，单果重168g，纵径6.7cm，横径6.8cm，倒卵圆形或近圆形，不规则；果皮绿黄色或黄色；果心中大，5心室；果肉白色，肉质细，松脆，汁液多，味甜；含可溶性固形物12.0%；品质上等或中上等。

24. Jinsuli

Origin and Distribution Jinsuli, originated from a cross between Yali (female parent) and Jinli (male parent) made by Fruit Research Institute, Shanxi Academy of Agricultural Sciences in 1957 and released in 1972, is grown in Shanxi, Shaanxi, Yunnan, etc.

Main Characters Tree: moderately vigorous, productive, early or medium precocity. Leaf: 10.4cm × 7.4cm in size, ovate. Initial leaf: red with less green. Flower: white bud tinged with pink on edge, edge of petal also pink, 5 to 7 flowers per cluster in average of 6.2; stamen number: 20 to 26, averaging 21.5; corolla diameter: 3.7cm. Fruit: matures in mid or late September in central Shanxi Province, 168g per fruit, 6.7cm long, 6.8cm wide, obovate or sub-round, irregular; greenish-yellow or yellow skin, medium core, 5 locules, flesh white, fine, crisp tender, juicy, sweet; TSS 12.0%; good or above medium in quality.

25. 金水 1 号

来源及分布 2n=34，湖北省农业科学院果树茶叶研究所育成，母本为长十郎，父本为江岛，1958 年杂交，1974 年通过鉴定。在湖北等地有栽培。

主要性状 树势强，丰产，始果年龄早。叶片长 14.3cm，宽 8.8cm，椭圆形，初展叶褐红色。花蕾白色，每花序 4 ~ 10 朵花，平均 5.8 朵；雄蕊 18 ~ 23 枚，平均 21.4 枚。在湖北武汉，果实 8 月下旬成熟，单果重 294g，纵径 7.5cm，横径 8.2cm，阔倒卵圆形或近圆形；果皮绿色；果心中大，5 心室；果肉白色，肉质较细，松脆，汁液多，味酸甜；含可溶性固形物 11.0%，可滴定酸 0.24%；品质中上等。果实不耐贮藏。

25. Jinshui No.1

Origin and Distribution Jinshui No.1 (2n=34), originated from a cross between Choujuurou (female parent) and Enoshima (male parent) made by Institute of Fruit and Tea, Hubei Academy of Agricultural Sciences in 1958 and released in 1974, is grown in Hubei, etc.

Main Characters Tree: vigorous, productive, early precocity. Leaf: 14.3cm × 8.8cm in size, elliptical. Initial leaf: brownish-red. Flower: white bud, 4 to 10 flowers per cluster in average of 5.8; stamen number: 18 to 23, averaging 21.4. Fruit: matures in late August in Wuhan, Hubei Province, 294g per fruit, 7.5cm long, 8.2cm wide, broadly obovate or sub-round, green skin, medium core, 5 locules, flesh white, relatively fine, crisp tender, juicy, sour-sweet; TSS 11.0%, TA 0.24%; quality above medium; storage life short.

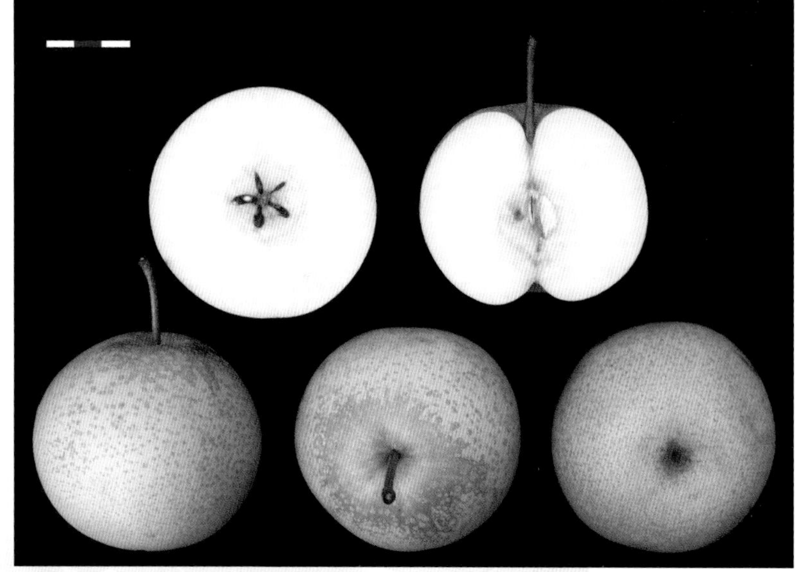

26. 金水 2 号

来源及分布 2n=34，又名翠伏，湖北省农业科学院果树茶叶研究所育成，母本为长十郎，父本为江岛。在湖北等地有栽培。

主要性状 树势强，丰产，始果年龄早。叶片长 11.8cm，宽 8.0cm，阔卵圆形，初展叶绿色，着淡红色。花蕾白色，边缘浅粉红色，每花序 6 ~ 8 朵花，平均 7.1 朵；雄蕊 20 ~ 28 枚，平均 24.1 枚。在湖北武汉，果实 7 月下旬成熟，单果重 183g，纵径 6.8cm，横径 7.1cm，倒卵圆形；果皮绿色或黄绿色；果心中大，5 心室；果肉乳白色，肉质细，松脆，汁液特多，味酸甜；含可溶性固形物 11.2%，可滴定酸 0.24%；品质上等。果实不耐贮藏。

26. Jinshui No.1

Origin and Distribution Jinshui No.1 (2n=34), also known as Cuifu, originated from a cross between Choujuurou (female parent) and Enoshima (male parent) made by Institute of Fruit and Tea, Hubei Academy of Agricultural Sciences, is grown in Hubei, etc.

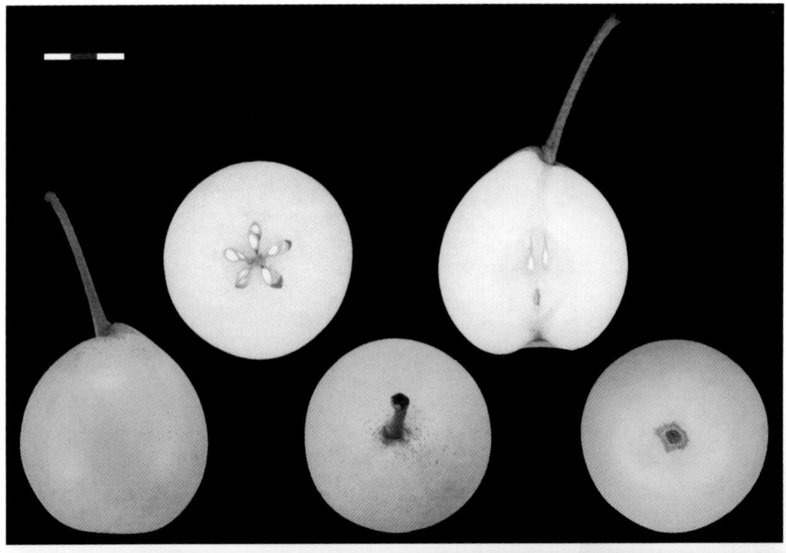

Main Characters Tree: vigorous, productive, early precocity. Leaf: 11.8cm × 8.0cm in size, broadly ovate. Initial leaf: green tinged with light red. Flower: white bud tinged with light pink on edge, 6 to 8 flowers per cluster in average of 7.1; stamen number: 20 to 28, averaging 24.1. Fruit: matures in late July in Wuhan, Hubei Province, 183g per fruit, 6.8cm long, 7.1cm wide, obovate, green or yellowish-green skin, medium core, 5 locules, flesh creamy white, fine, crisp tender, extremely juicy, sour-sweet; TSS 11.2%, TA 0.24%; quality good; storage life short.

27. 金水酥

来源及分布 2n=34，湖北省农业科学院果树茶叶研究所育成，母本为兴隆麻梨，父本为金水1号，1973年杂交，1985年通过鉴定。湖北等地有栽培。

主要性状 树势中庸，丰产，始果年龄早。叶片长10.0cm，宽6.2cm，卵圆形，初展叶绿色，微着红色。花蕾白色，边缘浅粉红色，每花序5～10朵花，平均7.3朵；雄蕊18～29枚，平均24.9枚。在湖北武汉，果实7月中旬成熟，单果重152g，纵径6.2cm，横径6.6cm，圆形或倒卵圆形；果皮绿色；果心小，5心室；果肉白色，肉质细，疏松，汁液特多，味酸甜；含可溶性固形物12.2%，可滴定酸0.40%；品质上等。果实不耐贮藏。

27. Jinshuisu

Origin and Distribution Jinshuisu (2n=34), originated from a cross between Xinglong Mali (female parent) and Jinshui No.1 (male parent) made by Institute of Fruit and Tea, Hubei Academy of Agricultural Sciences in 1973 and released in 1985, is grown in Hubei, etc.

Main Characters Tree: moderately vigorous, productive, early precocity. Leaf: 10.0cm × 6.2cm in size, ovate. Initial leaf: green with less red. Flower: white bud tinged with light pink on edge, 5 to 10 flowers per cluster in average of 7.3; stamen number: 18 to 29, averaging 24.9. Fruit: matures in mid July in Wuhan, Hubei Province, 152g per fruit, 6.2cm long, 6.6cm wide, round or obovate, green skin, small core, 5 locules, flesh white, fine, tender, extremely juicy, sour-sweet; TSS 12.2%, TA 0.40%; quality good; storage life short.

28. 龙园洋梨

来源及分布 黑龙江省农业科学院园艺分院育成，母本为龙香梨，父本为63-1-76和63-2-5的混合花粉，1978年杂交，2000年通过审定。在黑龙江省南部及东部、吉林等地有栽培。

主要性状 树势中庸，较丰产，始果年龄早。叶片长8.7cm，宽5.7cm，卵圆形，初展叶绿色，微着红色。花蕾白色，小花蕾边缘粉红色，每花序5～8朵花，平均6.8朵；雄蕊20～21枚，平均20.1枚；花冠直径3.7cm。在黑龙江哈尔滨，果实9月中旬成熟，单果重132g，纵径6.8cm，横径5.9cm，葫芦形或倒卵圆形；果皮黄色，阳面有红晕；果心中大，5心室；果肉白色，肉质细，脆，后熟变软，汁液中多，味甜酸，有香气；含可溶性固形物13.43%，可滴定酸0.54%；品质中上等。

28. Longyuan Yangli

Origin and Distribution Longyuan Yangli, originated from a cross between Longxiangli (female parent) and mixed pollen, 63-1-76/63-2-5 as male parent made by Horticultural Branch, Heilongjiang Academy of Agricultural Sciences in 1978 and released in 2000, is grown in eastern and southern parts of Heilongjiang, Jilin, etc.

Main Characters Tree: moderately vigorous, relatively productive, early precocity. Leaf: 8.7cm × 5.7cm in size, ovate. Initial leaf: green slightly tinged with red. Flower: white bud, small one tinged with pink on edge, 5 to 8 flowers per cluster in average of 6.8; stamen number: 20 to 21, averaging 20.1; corolla diameter: 3.7cm. Fruit: matures in mid-September in Harbin, Heilongjiang Province, 132g per fruit, 6.8cm long, 5.9cm wide, pyriform or obovate, yellow skin covered with red on the side exposed to the sun, medium core, 5 locules, flesh white, fine, crisp, turning to soft after storage, mid-juicy, sweet-sour, aromatic; TSS 13.43%, TA 0.54%; quality above medium.

29. 龙园洋红

来源及分布 2n=3x=51，黑龙江省农业科学院园艺分院育成，母本为56-5-20，父本为乔玛，1981年杂交，2005年通过审定。在黑龙江省南部及东部等地有栽培。

主要性状 树势强，丰产，始果年龄早。叶片长12.1cm，宽7.1cm，卵圆形，初展叶绿色。花蕾白色，边缘粉红色，每花序8～12朵花，花粉极少。在黑龙江哈尔滨，果实9月中旬成熟，单果重167g，纵径7.4cm，横径6.8cm，短粗颈葫芦形，不整齐；果皮黄色，有红晕；果心小，5心室；果肉黄白色，肉质细，软，汁液多，味甜酸，有香气；含可溶性固形物16.05%，可滴定酸0.81%；品质上等。

29. Longyuan Yanghong

Origin and Distribution Longyuan Yanghong (2n=3x=51), originated from a cross between 56-5-20 (female parent) and Тема (male parent) made by Horticultural Branch, Heilongjiang Academy of Agricultural Sciences in 1981 and released in 2005, is grown in eastern and southern parts of Heilongjiang, etc.

Main Characters Tree: vigorous, productive, precocity early. Leaf: 12.1cm × 7.1cm in size, ovate. Initial leaf: green. Flower: white bud tinged with pink on edge, 8 to 12 flowers per cluster, pollen less. Fruit: matures in mid-September in Harbin, Heilongjiang Province, 167g per fruit, 7.4cm long, 6.8cm wide, short obtuse-pyriform, irregular, yellow skin covered with red on the side exposed to the sun, small core, 5 locules, flesh yellowish-white, fine, soft, juicy, sweet-sour, aromatic; TSS 16.05%, TA 0.81%; quality good.

30. 美人酥

来源及分布 中国农业科学院郑州果树研究所与新西兰皇家园艺与食品研究所选育，母本为幸水，父本为火把梨，1989 年杂交，2008 年通过审定。在河南、河北、云南、甘肃等地区有栽培。

主要性状 树势中庸，丰产，始果年龄早。叶片长 10.8cm，宽 5.2cm，卵圆形，初展叶绿色。花蕾白色，每花序 9 ~ 10 朵花，雄蕊 25 ~ 34 枚。在河南郑州，果实 9 月中、下旬成熟，单果重 186g，纵径 6.5cm，横径 6.4cm，卵圆形；果皮绿黄色，阳面有鲜红晕；果心小，5 或 6 心室；果肉淡黄白色，肉质细，松脆，汁液多，味甜酸；可溶性固形物含量 12.4%；品质上等或中上等。

30. Meirensu

Origin and Distribution Meirensu, originated from a cross between Kousui (female parent) and Huobali (male parent) made by Zhengzhou Fruit Research Institute, Chinese Academy of Agricultural Sciences, Horticulture and Food Research Institute of New Zealand Limited in 1989 and released in 2008, is grown in Henan, Hebei, Yunnan, and Gansu, etc.

Main Characters Tree: moderately vigorous, productive, precocity early. Leaf: 10.8cm × 5.2cm in size, ovate. Initial leaf: green. Flower: white bud, 9 to 10 flowers per cluster; stamen number: 25 to 34. Fruit: matures in mid or late September in Zhengzhou, Henan Province, 186g per fruit, 6.5cm long, 6.4cm wide, ovate, greenish-yellow skin covered with bright red on the side exposed to the sun, small core, 5 or 6 locules, flesh pale yellowish-white, fine, crisp tender, juicy, sweet-sour; TSS 12.4%; good or above medium in quality.

31. 满天红

来源及分布 中国农业科学院郑州果树研究所与新西兰皇家园艺与食品研究所选育，母本为幸水，父本为火把梨，1989 年杂交，2008 年通过审定。在河南、河北、云南、甘肃等地区有栽培。

主要性状 树势强，丰产，始果年龄早。叶片长 10.4cm，宽 6.7cm，卵圆形，初展叶绿色，着红色，有茸毛。花蕾白色，每花序 8 ~ 9 朵花，平均 8.7 朵；雄蕊 29 ~ 37 枚，平均 31.5 枚；花冠直径 4.4cm。在河南郑州，果实 9 月下旬成熟，单果重 294g，近圆形；果皮绿黄色，阳面有红晕；果心较小，5 或 6 心室；果肉淡黄白色，肉质细，脆而紧密，汁液多，味甜酸或微酸；含可溶性固形物 12.9%；品质上等或中上等。果实较耐贮藏。

31. Mantianhong

Origin and Distribution Mantianhong, originated from a cross between Kousui (female parent) and Huobali (male parent) made by Zhengzhou Fruit Research Institute, Chinese Academy of Agricultural Sciences and Horticulture and Food Research Institute of New Zealand Limited in 1989 and released in 2008, is grown in Henan, Hebei, Yunnan, and Gansu, etc.

Main Characters Tree: vigorous, productive, precocity early. Leaf: 10.4cm × 6.7cm in size, ovate. Initial leaf: green tinged with red, pubescent. Flower: white bud, 8 to 9 flowers per cluster in average of 8.7; stamen number: 29 to 37, averaging 31.5; corolla diameter: 4.4cm. Fruit: matures in late September in Zhengzhou, Henan Province, 294g per fruit, sub-round, greenish-yellow skin covered with red on the side exposed to the sun, relatively small core, 5 or 6 locules, flesh pale yellowish-white, fine, juicy, dense and crisp, sweet-sour or sub-sour; TSS 12.9%; good or above medium in quality; storage life relatively long.

32. 清香

来源及分布 浙江省农业科学院园艺研究所育成，母本为新世纪，父本为三花，1978年杂交，2005年通过认定。在浙江、福建、江西等省有栽培。

主要性状 树势中庸，丰产，始果年龄早。叶片长11.1cm，宽7.0cm，卵圆形，初展叶绿色，着橘红色。花蕾白色，每花序5～9朵花，平均7.1朵；雄蕊20～33枚，平均28.5枚。在浙江杭州，果实8月上、中旬成熟，较翠冠晚20d左右，单果重210g，纵径6.1cm，横径6.7cm，圆形、长圆形或圆锥形；果皮黄褐色，或绿黄色着大面积果锈；果心极小，5心室；果肉白色，肉质较细，松脆，汁液多，味甜；含可溶性固形物11.5%；品质中上等。

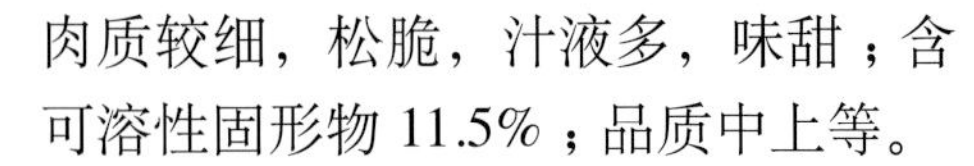

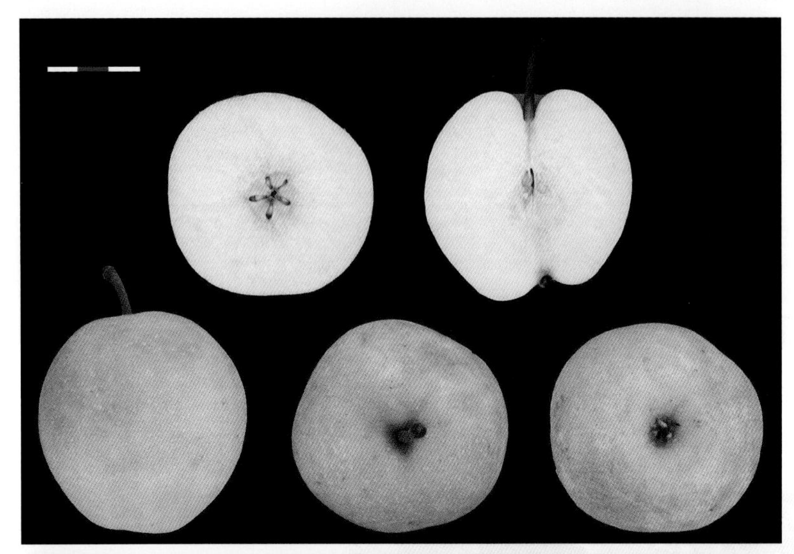

32. Qingxiang

Origin and Distribution Qingxiang, originated from a cross between Shinseiki (female parent) and Sanhua (male parent) made by the Horticultural Research Institute, Zhejiang Academy of Agricultural Sciences in 1978 and released in 2005, is grown in Zhejiang, Fujian, Jiangxi, etc.

Main Characters Tree: moderately vigorous, productive, precocity early. Leaf: 11.1cm × 7.0cm in size, ovate. Initial leaf: green tinged with orange red. Flower: white bud, 5 to 9 flowers per cluster in average of 7.1; stamen number: 20 to 33, averaging 28.5. Fruit: matures in early or mid August in Hangzhou, Zhejiang Province, 210g per fruit, 6.1cm long, 6.7cm wide, long round, round, or conical, russet skin or greenish-yellow skin covered with obvious russet, extremely small core, 5 locules, flesh white, relatively fine, crisp tender, juicy, sweet; TSS 11.5%; quality above medium.

33. 秦丰梨

来源及分布　原陕西省果树研究所育成，母本为茌梨，父本为象牙梨，1957 年杂交，1988 年通过审定。在陕西、山西、甘肃等地有栽培。

主要性状　树势强，丰产，始果年龄中等。叶片长 12.7cm，宽 7.1cm，卵圆形，初展叶褐红色，微显绿色。花蕾白色，边缘粉红色，每花序 4 ~ 6 朵花，平均 4.8 朵；雄蕊 20 ~ 21 枚，平均 20.2 枚；花冠直径 3.9cm。在陕西眉县，果实 9 月中旬成熟，单果重 273g，纵径 8.5cm，横径 8.0cm，椭圆形；果皮绿黄色或黄色；果心中大，5 心室；果肉白色，肉质较细，疏松，汁液多，味甜；含可溶性固形物 12.62%；品质中上等。

33. Qinfengli

Origin and Distribution Qinfengli, originated from a cross between Chili (female parent) and Xiangyali (male parent) made by former Shaanxi Fruit Research Institute in 1957 and released in 1988, is grown in Shaanxi, Shanxi, Gansu, etc.

Main Characters Tree: vigorous, productive, medium precocity. Leaf: 12.7cm × 7.1cm in size, ovate. Initial leaf: brownish-red with less green. Flower: white bud tinged with pink on edge, 4 to 6 flowers per cluster in average of 4.8; stamen number: 20 to 21, averaging 20.2; corolla diameter: 3.9cm. Fruit: matures in mid-September in Meixian, Shaanxi Province, 273g per fruit, 8.5cm long, 8.0cm wide, elliptical, greenish-yellow or yellow skin, medium core, 5 locules, flesh white, relatively fine, tender, juicy; TSS 12.62%; quality above medium.

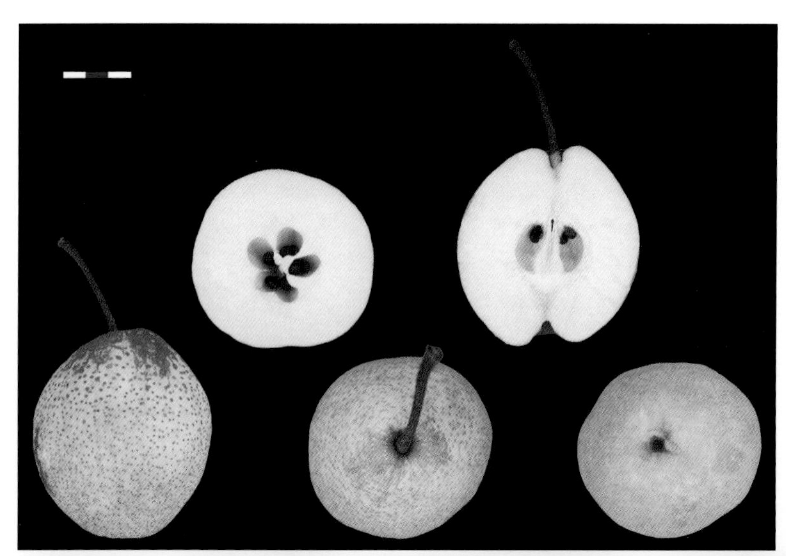

34. 秦酥梨

来源及分布 原陕西省果树研究所育成，母本为砀山酥梨，父本为黄县长把，1957 年杂交，1978 年定名。在甘肃、陕西、山西等地有栽培。

主要性状 树势强，丰产，始果年龄中等。叶片长 10.0cm，宽 8.6cm，圆形或阔卵圆形，初展叶绿色，着红色。花蕾白色，每花序 5 ～ 8 朵花，平均 6.3 朵；雄蕊 20 ～ 29 枚，平均 23.7 枚；花冠直径 4.1cm。在陕西杨凌，果实 10 月初成熟，单果重 232g，纵径 7.6cm，横径 7.4cm，圆柱形；果皮绿黄色或黄色，果肩部有锈；果心小，5 心室；果肉白色，肉质细，松脆，汁液多，味甜；含可溶性固形物 12.2%；品质上等。果实耐贮藏。

34. Qinsuli

Origin and Distribution Qinsuli, originated from a cross between Dangshan Suli (female parent) and Huangxian Changba (male parent) made by former Shaanxi Fruit Research Institute in 1957 and released in 1978, is grown in Gansu, Shaanxi, Shanxi, etc.

Main Characters Tree: vigorous, productive, medium precocity. Leaf: 10.0cm × 8.6cm in size, round or broadly ovate. Initial leaf: green tinged with red. Flower: white bud, 5 to 8 flowers per cluster in average of 6.3; stamen number: 20 to 29, averaging 23.7; corolla diameter: 4.1cm. Fruit: matures in early October in Yangling, Shaanxi Province, 232g per fruit, 7.6cm long, 7.4cm wide, cylindrical, greenish-yellow or yellow skin, russet on the stalk end, small core, 5 locules, flesh white, fine, crisp tender, juicy, sweet; TSS 12.2%; quality good; storage life long.

35. 秋香

来源及分布 原黑龙江省农业科学院园艺研究所育成，母本为59-89-1，父本为56-11-155，1971年杂交，1989年通过审定。在黑龙江、吉林、辽宁、内蒙古等地有栽培。

主要性状 树势中庸，丰产，始果年龄中等。叶片长8.7cm，宽5.2cm，卵圆形，初展叶绿色。花蕾白色，边缘浅粉红色，每花序9～11朵花，平均10.0朵；雄蕊17～20枚，平均19.1枚；花冠直径3.3cm。在黑龙江哈尔滨，果实9月初成熟，单果重65g，纵径4.3cm，横径4.7cm，圆形或卵圆形；果皮黄色；果心中大或大，5心室；果肉白色，肉质细，软溶，汁液多，味甜酸，有香气；含可溶性固形物17.8%，可滴定酸0.8%；品质中上等。

35. Qiuxiang

Origin and Distribution Qiuxiang, originated from a cross between 59-89-1 (female parent) and 56-11-155 (male parent) made by former Horticultural Research Institute, Heilongjiang Academy of Agricultural Sciences in 1971 and released in 1989, is grown in Heilongjiang, Jilin, Liaoning, and Inner Mongolia, etc.

Main Characters Tree: moderately vigorous, productive, medium precocity. Leaf: 8.7cm × 5.2cm in size, ovate. Initial leaf: green. Flower: white bud tinged with light pink on edge, 9 to 11 flowers per cluster in average of 10.0; stamen number: 17 to 20, averaging 19.1; corolla diameter: 3.3cm. Fruit: matures in early September in Harbin, Heilongjiang Province, 65g per fruit, 4.3cm long, 4.7cm wide, round or ovate, yellow skin, medium or large core, 5 locules, flesh white, fine, melting, juicy, sweet-sour, aromatic; TSS 17.8%, TA 0.8%; quality above medium.

36. 七月酥

来源及分布　中国农业科学院郑州果树研究所育成，母本为幸水，父本为早酥，1980年杂交，1999年通过审定。在华北平原、西北等地有栽培。

主要性状　树势较强，较丰产，始果年龄早。叶片长12.3cm，宽6.2cm，长椭圆形，初展叶绿色。花蕾白色，小花蕾边缘浅粉红色，每花序5～10朵花，平均7.3朵；雄蕊29～36枚，平均33.1枚；花冠直径3.3cm。在河南郑州，果实7月上旬成熟，单果重184g，纵径6.0cm，横径6.4cm，阔卵圆形或扁圆形；果皮黄绿色；果心小，6～8心室；果肉白色，肉质细，疏松，汁液多，味甜；含可溶性固形物12.5%；品质上等。不抗褐斑病和轮纹病，易早期落叶。

36. Qiyuesu

Origin and Distribution　Qiyuesu, originated from a cross between Kousui (female parent) and Zaosu (male parent) made by Zhengzhou Fruit Research Institute, Chinese Academy of Agricultural Sciences in 1980 and released in 1999, is grown in North China Plain and Northwest China, etc.

Main Characters　Tree: relatively vigorous, relatively productive, precocity early. Leaf: 12.3cm × 6.2cm in size, elongated elliptical. Initial leaf: green. Flower: white bud, small one tinged with light pink on edge, 5 to 10 flowers per cluster in average of 7.3; stamen number: 29 to 36, averaging 33.1; corolla diameter: 3.3cm. Fruit: matures in early July in Zhengzhou, Henan Province, 184g per fruit, 6.0cm long, 6.4cm wide, broadly ovate or oblate, yellowish-green skin, small core, 6 to 8 locules, flesh white, fine, tender, juicy, sweet; TSS 12.5%; quality good. Susceptible to brown spot and ring rot, leaf easily early deciduous.

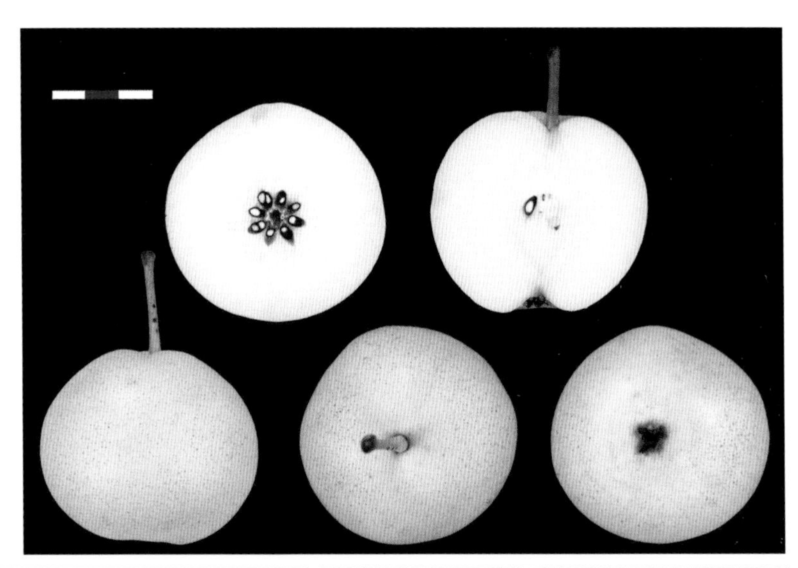

37. 沙 01

来源及分布 2n=4x=68，新疆巴音郭楞蒙古自治州沙依东园艺场和新疆农业科学院选育，为库尔勒香梨芽变，1969 年发现。在南疆有栽培。

主要性状 树势强，产量中等，始果年龄中等。叶片长 10.9cm，宽 7.3cm，阔卵圆形，初展叶绿色，微着红色。花蕾白色，边缘浅粉红色，每花序 5 ～ 8 朵花，平均 6.6 朵；雄蕊 21 ～ 30 枚，平均 26.3 枚；花冠直径 4.8cm。在新疆库尔勒，果实 9 月上旬成熟，单果重 156g，纵径 7.0cm，横径 6.9cm，阔卵圆形，不规则；果皮黄绿色，阳面有暗红色条纹；果心中大，5 心室；果肉白色，肉质细，松脆，汁液多，味甜，有香气；含可溶性固形物 11.38%；品质上等。

37. Sha 01

Origin and Distribution Sha 01 (2n=4x=68), a bud mutation of Korla Pear, which was found in 1969 by Shayidong Horticultural Farm, Bayingol Mongolian Autonomous Prefecture, and Xinjiang Academy of Agricultural Sciences, is grown in Southern Xinjiang.

Main Characters Tree: vigorous, medium productive, medium precocity. Leaf: 10.9cm × 7.3cm in size, broadly ovate. Initial leaf: green tinged with light red. Flower: white bud tinged with light pink on edge, 5 to 8 flowers per cluster in average of 6.6; stamen number: 21 to 30, averaging 26.3; corolla diameter: 4.8cm. Fruit: matures in early September in Korla, Xinjiang, 156g per fruit, 7.0cm long, 6.9cm wide, broadly ovate, irregular; yellowish-green skin covered with striped dark red on the side exposed to the sun, medium core, 5 locules, flesh white, fine, crisp tender, juicy, sweet, aromatic; TSS 11.38%; quality good.

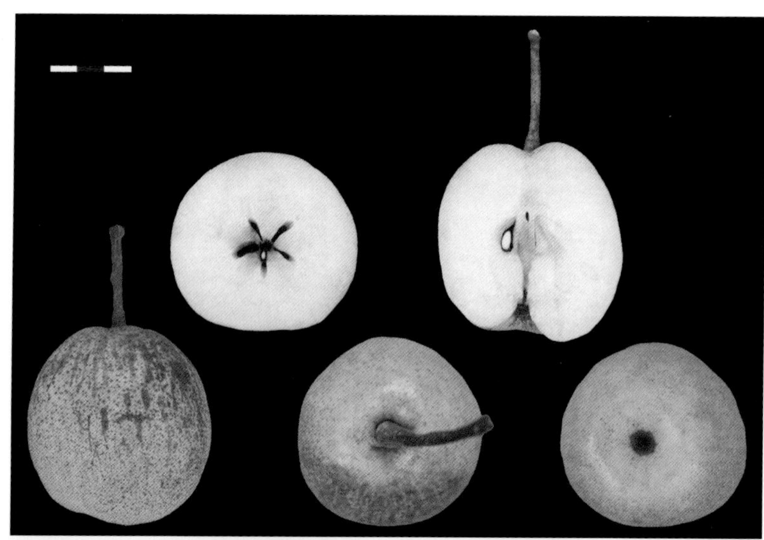

38. 硕丰

来源及分布 山西省农业科学院果树研究所育成，母本为苹果梨，父本为砀山酥梨，1972 年杂交，1995 年通过审定。在山西、内蒙古等地有栽培。

主要性状 树势较强，丰产，始果年龄早或中等。叶片长 11.1cm，宽 6.4cm，卵圆形，初展叶绿色，微着红色。花蕾白色，边缘淡粉红色，每花序 6 ～ 9 朵花，平均 7.9 朵；雄蕊 20 ～ 33 枚，平均 24.7 枚；花冠直径 4.4cm。在山西晋中地区，果实 9 月初成熟，单果重 250g，纵径 7.8cm，横径 8.5cm，近圆形或阔倒卵圆形，不规则；果皮绿黄色或黄色；果心小，5 心室；果肉白色，肉质细，松脆，汁液多，味甜或酸甜；含可溶性固形物 12.0%；品质上等。果实耐贮藏。

38. Shuofeng

Origin and Distribution Shuofeng, originated from a cross between Pingguoli (female parent) and Dangshan Suli (male parent) made by Fruit Research Institute, Shanxi Academy of Agricultural Sciences in 1972 and released in 1995, is grown in Shanxi, Inner Mongolia, etc.

Main Characters Tree: relatively vigorous, productive, early or medium precocity. Leaf: 11.1cm × 6.4cm in size, ovate. Initial leaf: green tinged with light red. Flower: white bud tinged with light pink on edge, 6 to 9 flowers per cluster in average of 7.9; stamen number: 20 to 33, averaging 24.7; corolla diameter: 4.4cm. Fruit: matures in early September in central Shanxi Province, 250g per fruit, 7.8cm long, 8.5cm wide, sub-round or broadly ovate, irregular, greenish-yellow or yellow skin, small core, 5 locules, flesh white, fine, crisp tender, juicy, sweet or sour-sweet; TSS 12.0%; quality good; storage life long.

39. 五九香

来源及分布 中国农业科学院果树研究所育成，母本为鸭梨，父本为巴梨，1952 年杂交，1959 年定名。在辽宁、北京、甘肃等地有栽培。

主要性状 树势较强，树冠紧凑，丰产稳产，始果年龄中等。叶片长 8.5cm，宽 5.6cm，卵圆形，初展叶绿色。花蕾白色，小花蕾边缘淡粉红色，每花序 5 ～ 6 朵花，平均 5.8 朵；雄蕊 19 ～ 23 枚，平均 20.8 枚；花冠直径 3.7cm。在辽宁兴城，果实 9 月上、中旬成熟，单果重 272g，纵径 11.7cm，横径 7.3cm，长粗颈葫芦形；果皮绿黄色，部分果实阳面有淡红晕，肩部有片锈；果心中大，5 心室；果肉白色，采收时肉质紧脆，经后熟变软，中粗，汁液中多，味酸甜；含可溶性固形物 12.20%；品质中上等。

39. Wujiuxiang

Origin and Distribution Wujiuxiang, originated from a cross between Yali (female parent) and Bartlett (male parent) made by Research Institute of Pomology, Chinese Academy of Agricultural Sciences in 1952 and released in 1959, is grown in Liaoning, Beijing, Gansu, etc.

Main Characters Tree: relatively vigorous, compact, productive, stable production, medium precocity. Leaf: 8.5cm × 5.6cm in size, ovate. Initial leaf: green. Flower: white bud, small one tinged with light pink on edge, 5 to 6 flowers per cluster in average of 5.8; stamen number: 19 to 23, averaging 20.8; corolla diameter: 3.7cm. Fruit: matures in early or mid September in Xingcheng, Liaoning Province, 272g per fruit, 11.7cm long, 7.3cm wide, long obtuse-pyriform, greenish-yellow skin, some covered with light red on the side exposed to the sun, russet on the stalk end, medium core, 5 locules, flesh white, dense and crisp when harvested, turning to soft after storage, mid-coarse, mid-juicy, sour-sweet; TSS 12.20%; quality above medium.

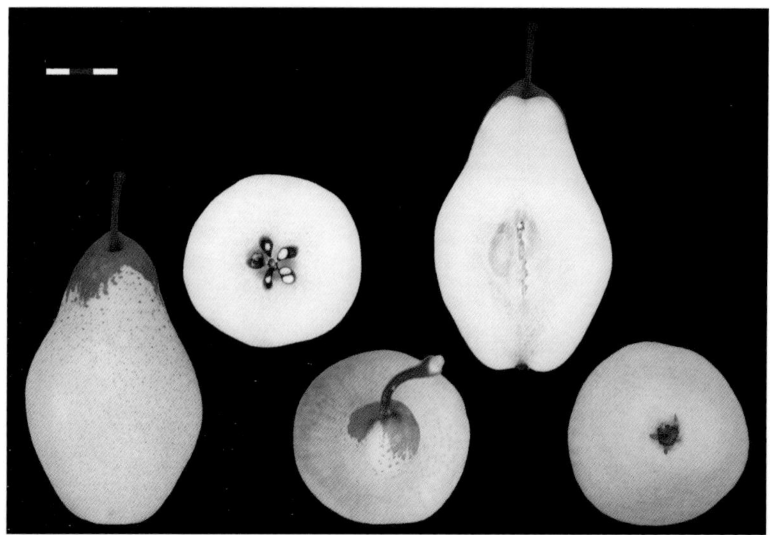

40. 新杭

来源及分布 原浙江农业大学育成，母本为新世纪，父本为杭青，1976年杂交，1990年选出。在浙江、江苏、福建等地有栽培。

主要性状 树势中庸，丰产，始果年龄早。叶片长12.9cm，宽8.4cm，卵圆形，初展叶红色，微显绿色，有茸毛。花蕾白色，每花序7～8朵花，平均7.6朵；雄蕊25～35枚，平均28.6枚；花冠直径4.3cm。在浙江杭州，果实7月中、下旬成熟，单果重250g，纵径7.4cm，横径7.0cm，圆形，外观美；果皮黄绿色；果心中大，5心室；果肉淡黄白色，肉质较细，松脆，汁液多，味酸甜；含可溶性固形物11.5%；品质中上等。

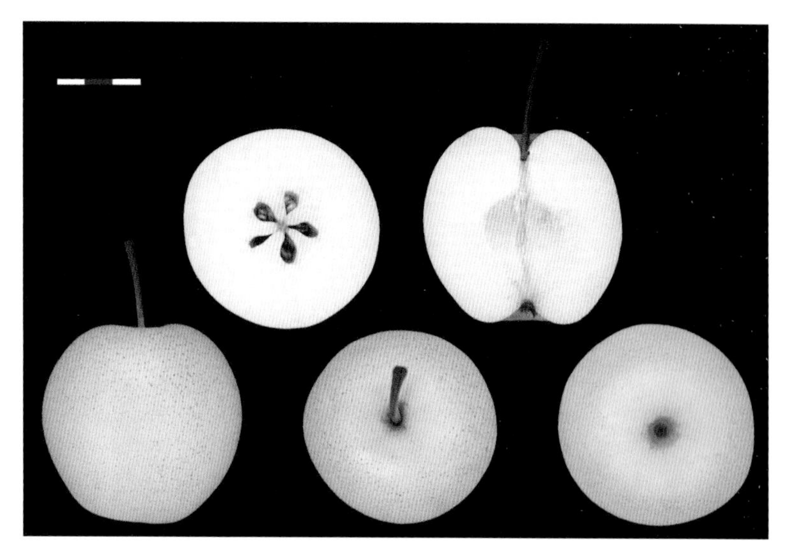

40. Xinhang

Origin and Distribution Xinhang, originated from a cross between Shinseiki (female parent) and Hangqing (male parent) made by former Zhejiang Agricultural University in 1976 and released in 1990, is grown in Zhejiang, Jiangsu, Fujian, etc.

Main Characters Tree: moderately vigorous, productive, early precocity. Leaf: 12.9cm × 8.4cm in size, ovate. Initial leaf: red with less green, pubescent. Flower: white bud, 7 to 8 flowers per cluster in average of 7.6; stamen number: 25 to 35, averaging 28.6; corolla diameter: 4.3cm. Fruit: matures in mid or late July in Hangzhou, Zhejiang Province, 250g per fruit, 7.4cm long, 7.0cm wide, round, surface beautiful, yellowish-green skin, medium core, 5 locules, flesh pale yellowish-white, relatively fine, crisp tender, juicy, sour-sweet; TSS 11.5%; quality above medium.

41. 新梨7号

来源及分布 塔里木大学育成，母本为库尔勒香梨，父本为早酥，2000年通过审定。在新疆、山东、河北等地有栽培。

主要性状 树势中庸，丰产，始果年龄早。叶片长10.3cm，宽6.5cm，椭圆形，初展叶绿色，叶尖带红色。花蕾白色，边缘粉红色，每花序7～9朵花，平均7.8朵；雄蕊21～30枚，平均25.7枚，花粉极少。在新疆阿拉尔，果实8月上旬成熟，单果重253g，纵径8.7cm，横径7.5cm，卵圆形、椭圆形或圆锥形，果面凹凸不平；果皮黄绿色或黄白色，阳面有红晕；果心中大，5心室；果肉黄白色，肉质疏松，极细，汁液多，味酸甜，含可溶性固形物11.88%；品质极上。果实耐贮藏。

41. Xinli No.7

Origin and Distribution Xinli No.7, originated from a cross between Korla Pear (female parent) and Zaosu (male parent) made by Tarim University and released in 2000, is grown in Xinjiang, Shandong, Hebei, etc.

Main Characters Tree: moderately vigorous, productive, early precocity. Leaf: 10.3cm × 6.5cm in size, elliptical. Initial leaf: green tinged with red on leaf apex. Flower: white bud tinged with pink on edge, 7 to 9 flowers per cluster in average of 7.8; stamen number: 21 to 30, averaging 25.7; pollen less. Fruit: matures in early August in Alear, Xinjiang, 253g per fruit, 8.7cm long, 7.5cm wide, ovate, elliptical, or conical, uneven, yellowish-green or yellowish-white skin covered with red on the side exposed to the sun, medium core, 5 locules, flesh yellowish-white, tender, extremely fine, juicy, sour-sweet; TSS 11.88%; quality extremely good; storage life long.

42. 西子绿

来源及分布 浙江大学育成，母本为新世纪，父本为八云 × 杭青，1976 年杂交，2001 年通过审定。在浙江、重庆等地有栽培。

主要性状 树势中庸，丰产，始果年龄早。叶片长 10.2cm，宽 7.8cm，卵圆形，初展叶绿色，着红色。花蕾白色，每花序 5 ~ 8 朵花，平均 6.7 朵，雄蕊 22 ~ 27 枚，平均 24.6 枚；花冠直径 3.9cm。在浙江杭州，果实 7 月底或 8 月初成熟，单果重 190g，纵径 6.0cm，横径 6.5cm，扁圆形，外观极美；果皮浅绿色或黄色；果心中大，5 心室；果肉白色，肉质细，松脆或疏松，汁液多，味淡甜；含可溶性固形物 11.39%；品质中上等。

42. Xizilü

Origin and Distribution Xizilü, originated from a cross between Shinseiki (female parent) and Yakumo × Hangqing (male parent) made by Zhejiang University in 1976 and released in 2001, is grown in Zhejiang, Chongqing, etc.

Main Characters Tree: moderately vigorous, productive, early precocity. Leaf: 10.2cm × 7.8cm in size, ovate. Initial leaf: green tinged with red. Flower: white bud, 5 to 8 flowers per cluster in average of 6.7; stamen number: 22 to 27, averaging 24.6; corolla diameter: 3.9cm. Fruit: matures in late July or early August in Hangzhou, Zhejiang Province, 190g per fruit, 6.0cm long, 6.5cm wide, oblate, surface very beautiful, light green or yellow skin, medium core, 5 locules, flesh white, fine, crisp tender or tender, juicy, light sweet; TSS 11.39%; quality above medium.

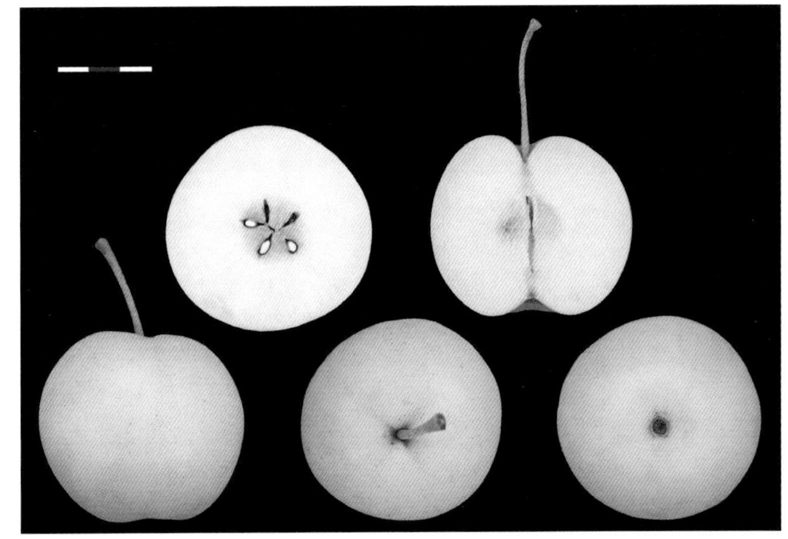

43. 雪青

来源及分布 浙江大学育成，母本为雪花梨，父本为新世纪，1977 年杂交，2001 年通过审定。在浙江、江西、湖北、四川等地有栽培。

主要性状 树势强，丰产，始果年龄早。叶片长 13.1cm，宽 8.1cm，阔卵圆形，初展叶绿色，着红色。花蕾白色，每花序 7 ~ 9 朵花，平均 8.1 朵；雄蕊 22 ~ 29 枚，平均 24.1 枚；花冠直径 4.3cm。在北京，果实 8 月上旬成熟，单果重 307g，纵径 7.3cm，横径 8.5cm，近圆形；果皮绿黄色；果心小，5 心室；果肉白色，肉质较细，松脆或疏松，汁液多，味甜或淡甜；含可溶性固形物 11.39%，可滴定酸 0.14%；品质上等或中上等。

43. Xueqing

Origin and Distribution Xueqing, originated from a cross between Xuehuali (female parent) and Shinseiki (male parent) made by Zhejiang University in 1977 and released in 2001, is grown in Zhejiang, Jiangxi, Hubei, Sichuan, etc.

Main Characters Tree: vigorous, productive, early precocity. Leaf: 13.1cm × 8.1cm in size, broadly ovate. Initial leaf: green tinged with red. Flower: white bud, 7 to 9 flowers per cluster in average of 8.1; stamen number: 22 to 29, averaging 24.1; corolla diameter: 4.3cm. Fruit: matures in early August in Beijing, 307g per fruit, 7.3cm long, 8.5cm wide, subround, greenish-yellow skin, small core, 5 locules, flesh white, relatively fine, crisp tender or tender, juicy, sweet or light sweet; TSS 11.39%, TA 0.14%; good or above medium in quality.

44. 玉露香

来源及分布　山西省农业科学院果树研究所育成，母本为库尔勒香梨，父本为雪花梨，1974年杂交，2003年通过审定。在山西、新疆、河北、陕西等地有栽培。

主要性状　树势中庸，丰产，始果年龄中等。叶片长11.1cm，宽7.3cm，卵圆形，初展叶绿色，着红色。花蕾白色，每花序5～7朵花，平均6.0朵；雄蕊20～23枚，平均20.5枚；花冠直径4.7cm。在山西晋中地区，果实8月下旬或9月上旬成熟，单果重294g，纵径8.1cm，横径8.3cm，近圆形或卵圆形，不规则，果面有沟；果皮绿黄色，果面有暗红色条红；果心小或中大，5心室；果肉白色或浅绿白色，肉质细，疏松，汁液多，味甜；含可溶性固形物13.9%；品质上等。果实较耐贮藏。

44. Yuluxiang

Origin and Distribution　Yuluxiang, was originated from a cross between Korla Pear (female parent) and Xuehuali (male parent) made by Fruit Research Institute, Shanxi Academy of Agricultural Sciences in 1974 and released in 2003, is grown in Shanxi, Xinjiang, Hebei, Shaanxi, etc.

Main Characters　Tree: moderately vigorous, productive, medium precocity. Leaf: 11.1cm × 7.3cm in size, ovate. Initial leaf: green tinged with red. Flower: white bud, 5 to 7 flowers per cluster in average of 6.0; stamen number: 20 to 23, averaging 20.5; corolla diameter: 4.7cm. Fruit: matures in late August or early September in central Shanxi Province, 294g per fruit, 8.1cm long, 8.3cm wide, sub-round or ovate, irregular, surface grooved, greenish-yellow skin covered with striped dark red on the side exposed to the sun, small or medium core, 5 locules, flesh white or pale greenish-white, fine, tender, juicy, sweet; TSS 13.9%; quality good; storage life relatively long.

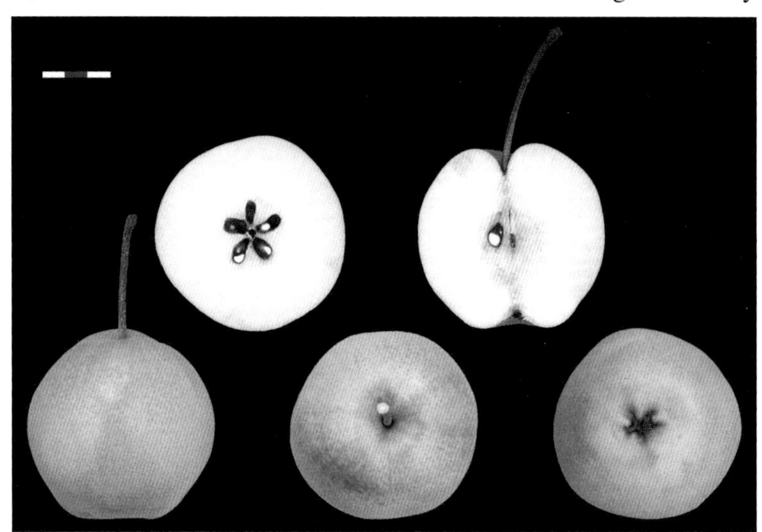

45. 玉酥梨

来源及分布　山西省农业科学院果树研究所育成，母本为砀山酥梨，父本为猪嘴梨，1972 年杂交，2009 年通过审定。在山西忻州、祁县、隰县等地有栽培。

主要性状　树势中庸，丰产，始果年龄中等。叶片长 10.4cm，宽 7.8cm，卵圆形，初展叶红色，微显绿色。花蕾白色，小花蕾边缘淡粉红色，每花序 6 ~ 8 朵花，平均 6.6 朵；雄蕊 20 ~ 29 枚，平均 24.2 枚；花冠直径 4.5cm。花粉量少。在山西晋中地区，果实 9 月下旬成熟，单果重 348g，纵径 10.2cm，横径 8.7cm，长圆形或长卵圆形；果皮黄绿色或黄色；果心小，3 ~ 5 心室；果肉白色，肉质细，疏松，汁液多，味甜；含可溶性固形物 11.0%；品质上等。果实耐贮藏。

45. Yusuli

Origin and Distribution Yusuli, originated from a cross between Dangshan Suli (female parent) and Zhuzuili (male parent) made by Fruit Research Institute, Shanxi Academy of Agricultural Sciences in 1972 and released in 2009, is grown in Xinzhou, Qixian, and Xixian of Shanxi Province, etc.

Main Characters Tree: moderately vigorous, productive, medium precocity. Leaf: 10.4cm × 7.8cm in size, ovate. Initial leaf: red with less green. Flower: white bud, small one tinged with light pink on edge, 6 to 8 flowers per cluster in average of 6.6; stamen number: 20 to 29, averaging 24.2; corolla diameter: 4.5cm, pollen less. Fruit: matures in late September in central Shanxi Province, 348g per fruit, 10.2cm long, 8.7cm wide, long round or long ovate, yellowish-green or yellow skin, small core, 3 to 5 locules, flesh white, fine, tender, juicy, sweet; TSS 11.0%; quality good; storage life long.

46. 早酥

来源及分布 中国农业科学院果树研究所育成，母本为苹果梨，父本为身不知，1956年杂交，1969年命名。在辽宁、河北、北京、江苏、甘肃、山西、陕西等地有栽培。

主要性状 树势强，丰产，始果年龄早。叶片长10.6cm，宽6.3cm，卵圆形，初展叶绿色，着红色，有茸毛。花蕾白色，边缘粉红色，每花序4～10朵花，平均7.8朵；雄蕊16～23枚，平均20.5枚；花冠直径3.9cm。在辽宁兴城，果实8月中、下旬成熟，单果重270g，纵径8.3cm，横径7.9cm，卵圆形或圆锥形；果皮绿色或黄绿色；果心中大，5心室；果肉白色，肉质细，松脆，汁液特多，味淡甜；含可溶性固形物11.15%；品质上等。

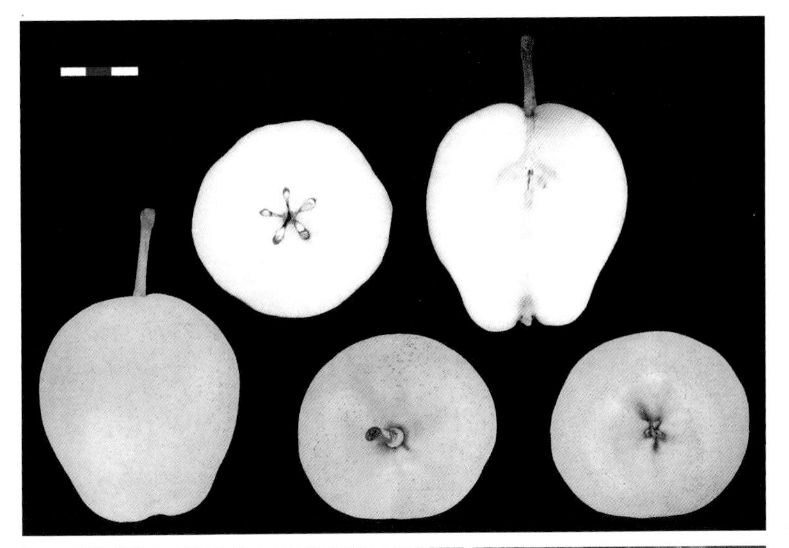

46. Zaosu

Origin and Distribution Zaosu, originated from a cross between Pingguoli (female parent) and Mishirazu (male parent) made by Research Institute of Pomology, Chinese Academy of Agricultural Sciences in 1956 and released in 1969, is grown in Liaoning, Hebei, Beijing, Jiangsu, Gansu, Shanxi, and Shaanxi, etc.

Main Characters Tree: vigorous, productive, early precocity. Leaf: 10.6cm × 6.3cm in size, ovate. Initial leaf: green tinged with red, pubescent. Flower: white bud tinged with pink on edge, 4 to 10 flowers per cluster in average of 7.8; stamen number: 16 to 23, averaging 20.5; corolla diameter: 3.9cm. Fruit: matures in mid or late August in Xingcheng, Liaoning Province, 270g per fruit, 8.3cm long, 7.9cm wide, ovate or conical, green or yellowish-green skin, medium core, 5 locules, flesh white, fine, very juicy, crisp tender, light sweet; TSS 11.15%; quality good.

47. 早冠

来源及分布　河北省农林科学院石家庄果树研究所育成，母本为鸭梨，父本为青云，1977 年杂交，2005 年通过审定。在河北、江苏、浙江等地有栽培。

主要性状　树势强，较丰产，始果年龄早，自交亲和。叶片长 10.4cm，宽 8.2cm，椭圆形，初展叶红色，带有绿色。花蕾白色，边缘淡粉红色，每花序 7 ~ 9 朵花，平均 8.0 朵；雄蕊 21 ~ 26 枚，平均 24.0 枚。在河北石家庄，果实 7 月下旬或 8 月上旬成熟，单果重 276g，纵径 8.1cm，横径 8.0cm，近圆形，果面平滑；果皮绿黄色或淡黄色；果心小，5 心室；果肉白色，肉质细，松脆，汁液多，味酸甜；含可溶性固形物 10.06%；品质上等。

47. Zaoguan

Origin and Distribution Zaoguan, originated from a cross between Yali (female parent) and Qingyun (male parent) made by Shijiazhuang Fruit Research Institute, Hebei Academy of Agro-Forestry Sciences in 1977 and released in 2005, is grown in Hebei, Jiangsu, Zhejiang, etc.

Main Characters Tree: vigorous, relatively productive, early precocity, self-fertile. Leaf: 10.4cm × 8.2cm in size, elliptical. Initial leaf: red with green. Flower: white bud tinged with light pink on edge, 7 to 9 flowers per cluster in average of 8.0; stamen number: 21 to 26, averaging 24.0. Fruit: matures in late July or early August in Shijiazhuang, Hebei Province, 276g per fruit, 8.1cm long, 8.0cm wide, sub-round, surface smooth, greenish-yellow or light yellow skin, small core, 5 locules, flesh white, fine, juicy, crisp tender, sour-sweet; TSS 10.06%; quality good.

48. 早美酥

来源及分布 中国农业科学院郑州果树研究所育成，母本为新世纪，父本为早酥，1982 年杂交，1998 年通过审定。在河南、安徽等地有栽培。

主要性状 树势较强，丰产，始果年龄早。叶片长 11.5cm，宽 6.7cm，椭圆形，初展叶绿色，有茸毛。花蕾白色，每花序 5 ~ 8 朵花，平均 6.1 朵；雄蕊 25 ~ 32 枚，平均 27.5 枚；花冠直径 3.4cm。在河南郑州，果实 7 月中旬成熟，单果重 250g，圆形；果皮绿黄色；果心中大，5 或 6 心室；果肉白色，肉质细，松脆，汁液多，味淡甜；含可溶性固形物 11.0%；品质中上等。果实不耐贮藏。

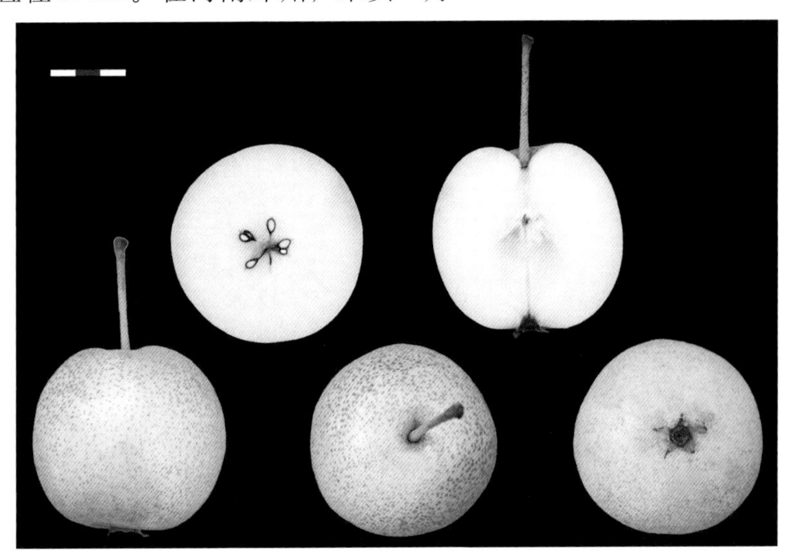

48. Zaomeisu

Origin and Distribution Zaomeisu, originated from a cross between Shinseiki (female parent) and Zaosu (male parent) made by Zhengzhou Fruit Research Institute, Chinese Academy of Agricultural Sciences in 1982 and released in 1998, is grown in Henan, Anhui, etc.

Main Characters Tree: relatively vigorous, productive, early precocity. Leaf: 11.5cm × 6.7cm in size, elliptical. Initial leaf: green, pubescent. Flower: white bud, 5 to 8 flowers per cluster in average of 6.1; stamen number: 25 to 32, averaging 27.5; corolla diameter: 3.4cm. Fruit: matures in mid-July in Zhengzhou, Henan Province, 250g per fruit, round, greenish-yellow skin, medium core, 5 or 6 locules, flesh white, fine, crisp tender, juicy, light sweet; TSS 11.0%; quality above medium; storage life short.

49. 蔗梨

来源及分布　吉林省农业科学院果树研究所育成，母本为苹果梨，父本为杭青，1977年杂交，2000年通过审定。在辽宁、吉林等地有栽培。

主要性状　树势强，丰产，始果年龄早。叶片长13.6cm，宽9.1cm，阔卵圆形，初展叶绿色，着红色。花蕾白色，边缘浅粉红色，每花序4～7朵花，平均6.0朵；雄蕊14～20枚，平均18.0枚；花冠直径3.8cm。在吉林公主岭，果实9月下旬成熟，单果重275g，纵径7.3cm，横径8.0cm，近圆形或圆锥形；果皮绿色，贮藏后转黄色；果心小，5心室；果肉白色，肉质细，脆，汁液多，味酸甜；含可溶性固形物11.25%；品质上等。

49. Zheli

Origin and Distribution Zheli, originated from a cross between Pingguoli (female parent) and Hangqing (male parent) made by Fruit Research Institute, Jilin Academy of Agricultural Sciences in 1977 and released in 2000, is grown in Liaoning, Jilin, etc.

Main Characters Tree: vigorous, productive, early precocity. Leaf: 13.6cm × 9.1cm in size, broadly ovate. Initial leaf: green tinged with red. Flower: white bud tinged with light pink on edge, 4 to 7 flowers per cluster in average of 6.0; stamen number: 14 to 20, averaging 18.0; corolla diameter: 3.8cm. Fruit: matures in late September in Gongzhuling, Jilin Province, 275g per fruit, 7.3cm long, 8.0cm wide, sub-round or conical, green skin, turning to yellow after storage, small core, 5 locules, flesh white, fine, crisp, juicy, sour-sweet; TSS 11.25%; quality good.

50. 珍珠梨

来源及分布 上海市农业科学院园艺研究所育成，母本为八云，父本为伏茄，1979 年杂交，1988 年育成。在浙江、上海等地有少量栽培。

主要性状 树势强，丰产，始果年龄早。叶片长 9.0cm，宽 5.1cm，卵圆形，初展叶绿色，微显红色。花蕾白色，小花蕾边缘微显红色；每花序 5 ~ 6 朵花，平均 5.8 朵；雄蕊 25 ~ 30 枚，平均 27.7 枚；花冠直径 3.7cm。在上海，果实 6 月下旬成熟，单果重 68g，纵径 5.4cm，横径 4.9cm，短葫芦形；果皮亮黄色；果心中大，5 心室；果肉淡黄白色，肉质细，松脆，汁液多，贮藏后汁液变少，味酸甜，微有香气；含可溶性固形物 10.70%；品质上等。果实不耐贮藏。

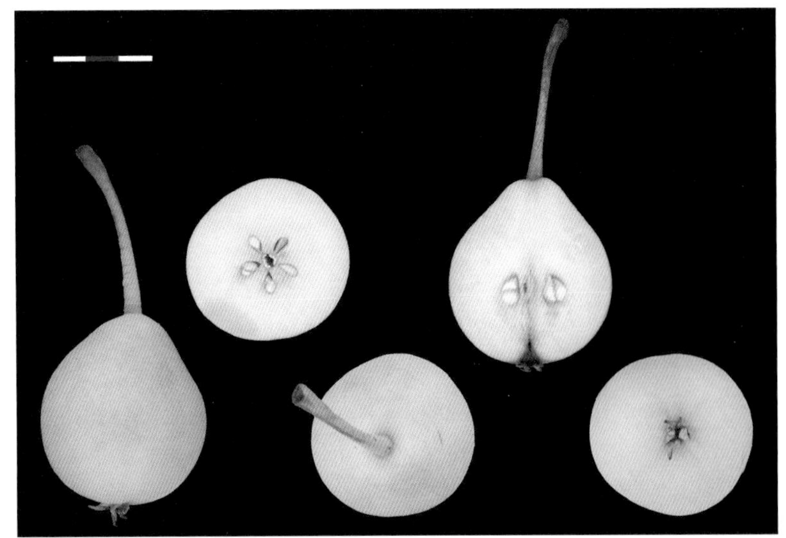

50. Zhenzhuli

Origin and Distribution Zhenzhuli, originated from a cross between Yakumo (female parent) and Beurré Giffard (male parent) made by Horticultural Research Institute, Shanghai Academy of Agricultural Sciences in 1979 and released in 1988, is grown limitedly in Zhejiang and Shanghai, etc.

Main Characters Tree: vigorous, productive, early precocity. Leaf: 9.0cm × 5.1cm in size, ovate. Initial leaf: green tinged with light red. Flower: white bud, small one slightly tinged with red on edge, 5 to 6 flowers per cluster in average of 5.8; stamen number: 25 to 30, averaging 27.7; corolla diameter: 3.7cm. Fruit: matures in late June in Shanghai, 68g per fruit, 5.4cm long, 4.9cm wide, short pyriform, brightly yellow skin, medium core, 5 locules, flesh pale yellowish-white, fine, crisp tender, juicy, turning to less juicy after storage, sour-sweet, slight aromatic; TSS 10.70%; quality good; storage life short.

51. 中梨 1 号

来源及分布 又名绿宝石，中国农业科学院郑州果树研究所育成，母本为新世纪，父本为早酥，1982 年杂交，2003 年通过审定。在华北、西南和长江中下游地区有栽培。

主要性状 树势较强，丰产，始果年龄早。叶片长 11.0cm，宽 6.3cm，椭圆形，初展叶绿色，有茸毛。花蕾白色，每花序 7 ~ 9 朵花，平均 8.5 朵；雄蕊 25 ~ 33 枚，平均 29.3 枚；花冠直径 3.4cm。在河南郑州，果实 7 月中旬成熟，单果重 264g，纵径 6.9cm，横径 8.1cm，近圆形或扁圆形；果皮绿色或绿黄色；果心中大，5 ~ 7 心室；果肉白色，肉质细，松脆或疏松，汁液多，味甜；含可溶性固形物 12.5%；品质上等。果实不耐贮藏。

51. Zhongli No.1

Origin and Distribution Zhongli No.1, known as Lübaoshi, originated from a cross between Shinseiki (female parent) and Zaosu (male parent) made by Zhengzhou Fruit Research Institute, Chinese Academy of Agricultural Sciences in 1982 and released in 2003, is grown in North China, Southwest China, and along middle and lower reaches of Yangtze River, etc.

Main Characters Tree: relatively vigorous, productive, early precocity. Leaf: 11.0cm × 6.3cm in size, elliptical. Initial leaf: green, pubescent. Flower: white bud, 7 to 9 flowers per cluster in average of 8.5; stamen number: 25 to 33, averaging 29.3; corolla diameter: 3.4cm. Fruit: matures in mid July in Zhengzhou, Henan Province, 264g per fruit, 6.9cm long, 8.1cm wide, sub-round or oblate, green or greenish-yellow skin, medium core, 5 to 7 locules, flesh white, fine, crisp tender or tender, juicy, sweet; TSS 12.5%; quality good; storage life short.

52. 早金香

来源及分布　中国农业科学院果树研究所育成，母本为矮香，父本为三季梨，1986 年杂交，2009 年通过审定。在辽宁、北京等地有栽培。

主要性状　树势中庸，丰产，始果年龄早。叶片卵圆形，初展叶绿色。花蕾白色，小花蕾边缘淡粉红色，每花序 5 ～ 7 朵花，平均 5.9 朵；雄蕊 22 ～ 31 枚，平均 28.2 枚；花冠直径 3.9cm。在辽宁兴城，果实 8 月中旬成熟，单果重 294g，纵径 9.7cm，横径 7.9cm，粗颈葫芦形；果皮黄绿色转黄色；果心小，5 心室；果肉白色，肉质细，软，汁液多，味酸甜，有香气；含可溶性固形物 13.50%；品质上等。

52. Zaojinxiang

Origin and Distribution Zaojinxiang, originated from a cross between Aixiang (female parent) and Docteur Jules Guyot (male parent) made by Research Institute of Pomology, Chinese Academy of Agricultural Sciences in 1986 and released in 2009, is grown in Liaoning, Beijing, etc.

Main Characters Tree: moderately vigorous, productive, early precocity. Leaf: ovate. Initial leaf: green. Flower: white bud, small one tinged with light pink on edge, 5 to 7 flowers per cluster in average of 5.9; stamen number: 22 to 31, averaging 28.2; corolla diameter: 3.9cm. Fruit: matures in mid August in Xingcheng, Liaoning Province, 294g per fruit, 9.7cm long, 7.9cm wide, obtuse-pyriform, yellowish-green skin, turning to yellow after storage, small core, 5 locules, flesh white, fine, soft, juicy, sour-sweet, aromatic; TSS 13.50%; quality good.

三、日本砂梨品种

Japanese Sand Pear Varieties

1. 爱宕

来源及分布 2n=34，日本冈山县龙井种苗株式会社育成，母本为二十世纪，父本为今村秋，1982 年命名。在河南、山东、辽宁等地有栽培。

主要性状 树势强，丰产，始果年龄早。叶片长 10.8cm，宽 6.7cm，卵圆形，初展叶暗红色。花蕾白色，边缘浅粉红色，每花序 3 ~ 8 朵花，平均 5.3 朵；雄蕊 18 ~ 22 枚，平均 20.3 枚。在辽宁熊岳，果实 10 月中、下旬成熟，单果重 415g，扁圆形；果皮黄褐色；果心小，5 心室；果肉白色，肉质细，脆，汁液多，味甜；含可溶性固形物 12.7%，可滴定酸 0.13%；品质中上等或上等。果实耐贮藏。

1. Atago

Origin and Distribution Atago (2n=34), originated from a cross between Nijisseiki (female parent) and Imamuraaki (male parent) made by Longjing Seedling Cooperation, Okayama Prefecture, Japan and released in 1982, is grown in Henan, Shandong, and Liaoning, etc.

Main Characters Tree: vigorous, productive, precocity early. Leaf: 10.8cm × 6.7cm in size, ovate. Initial leaf: dark red. Flower: white bud tinged with light pink on edge, 3 to 8 flowers per cluster in average of 5.3; stamen number: 18 to 22, averaging 20.3. Fruit: matures in mid or late October in Xiongyue, Liaoning Province, 415g per fruit, oblate, russet skin, small core, 5 locules, flesh white, fine, crisp, juicy, sweet; TSS 12.7%, TA 0.13%; good or above medium in quality; storage life long.

2. 丰水

来源及分布　2n=34，日本农林省园艺试验场1972年育成，母本为幸水，父本为I-33。在我国长江流域、黄河故道、华北平原等地有栽培。

主要性状　树势中庸，丰产，始果年龄早。叶片长12.8cm，宽8.4cm，卵圆形，初展叶红色，微显绿色，有茸毛。花蕾白色，每花序6～9朵花，平均7.3朵；雄蕊23～30枚，平均25.4枚；花冠直径4.1cm。在山东冠县，果实8月下旬成熟，单果重326g，纵径7.5cm，横径8.8cm，圆形或扁圆形；果皮锈褐色，果面用手触摸稍显粗糙，有棱沟；果心中大或小，5心室；果肉淡黄白色，肉质细，松脆或疏松，汁液极多或多，味甜；含可溶性固形物13.06%，可滴定酸0.13%；品质上等或极上。

2. Housui

Origin and Distribution Housui (2n=34), originated from a cross between Kosui (female parent) and I-33 (male parent), which was selected by Horticultural Experiment Station, the Ministry of Agriculture, Forestry of Japan and released in 1972, is grown mainly along Yangtze River basin, ancient riverbed of Yellow River, and North China plain, etc.

Main Characters Tree: moderately vigorous, productive, early precocity. Leaf: 12.8cm × 8.4cm in size, ovate. Initial leaf: red with less green, pubescent. Flower: white bud, 6 to 9 flowers per cluster in average of 7.3; stamen number: 23 to 30, averaging 25.4; corolla diameter: 4.1cm. Fruit: matures in late August in Guanxian, Shandong Province, 326g per fruit, 7.5cm long, 8.8cm wide, round or oblate, russet skin, somewhat rough when touched by fingers, grooved, medium or small core, flesh pale yellowish-white, fine, crisp tender or tender, juicy or very juicy, sweet; TSS 13.06%, TA 0.13%; good or extremely good in quality.

3. 菊水

来源及分布 2n=34，日本神奈川县菊池秋雄育成，母本为太白，父本为二十世纪。在我国江苏、江西、湖南、福建等地有少量栽培。

主要性状 树势中庸，丰产，始果年龄早。叶片长11.5cm，宽8.1cm，卵圆形，初展叶红色，有茸毛。花蕾白色，每花序8～11朵花，平均8.9朵；雄蕊26～33枚，平均29.4枚；花冠直径4.3cm。在湖北武汉，果实8月中旬成熟，单果重175g，扁圆形，不整齐；果皮黄绿色，有锈，不光滑；果心中大，5心室；果肉黄白色，肉质细，松软，汁液特多，味甜；含可溶性固形物14.0%，可滴定酸0.18%；品质上等或中上等。

3. Kikusui

Origin and Distribution Kikusui (2n=34), originated from a cross between Tahihaku (female parent) and Nijisseiki (male parent) by Kikuchi in Kanagawa Prefecture, Japan, is grown limitedly in Jiangsu, Jiangxi, Hunan, and Fujian, etc.

Main Characters Tree: moderately vigorous, productive, early precocity. Leaf: 11.5cm × 8.1cm in size, ovate. Initial leaf: red, pubescent. Flower: white bud, 8 to 11 flowers per cluster in average of 8.9; stamen number: 26 to 33, averaging 29.4; corolla diameter: 4.3cm. Fruit: matures in mid August in Wuhan, Hubei Province, 175g per fruit, oblate, irregular, greenish-yellow skin, russet, uneven, medium core, 5 locules, flesh yellowish-white, fine, tender and soft, extremely juicy, sweet; TSS 14.0%, TA 0.18%; good or above medium in quality.

4. 幸水

来源及分布 2n=34，产于日本静冈县，母本为菊水，父本为早生幸藏。在我国江苏、上海、江西等地有少量栽培。

主要性状 树势中庸，较丰产，始果年龄早或中等。叶片长12.8cm，宽8.0cm，卵圆形，初展叶红色，微显绿色，有茸毛。花蕾白色，每花序7～9朵花，平均8.2朵；雄蕊30～37枚，平均33.3枚；花冠直径4.7cm；花瓣数6～11枚。在湖北武汉，果实8月上旬成熟，单果重195g，扁圆形；果皮绿黄色，有果锈，或淡黄褐色；果心小，6～8心室；果肉白色，肉质细，松软，汁液多，味甜；含可溶性固形物12.3%；品质上等。

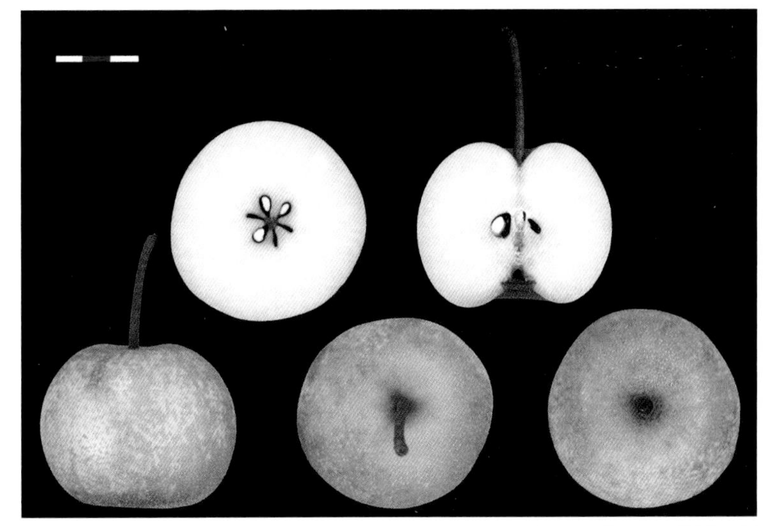

4. Kousui

Origin and Distribution Kousui (2n=34), originated from a cross between Kikusui (female parent) and Wasekouzou (male parent) made in Shizuoka Prefecture, Japan, is grown limitedly in Jiangsu, Shanghai, Jiangxi, etc.

Main Characters Tree: moderately vigorous, relatively productive, early or medium precocity. Leaf: 12.8cm × 8.0cm in size, ovate. Initial leaf: red with less green, pubescent. Flower: white bud, 7 to 9 flowers per cluster in average of 8.2; stamen number: 30 to 37, averaging 33.3; corolla diameter: 4.7cm. Petals: 6 to 11. Fruit: matures in early August in Wuhan, Hubei Province, 195g per fruit, oblate, greenish-yellow skin with russet or light brown, small core, 6 to 8 locules, flesh white, fine, tender, juicy, sweet; TSS 12.3%; quality good.

5. 新高

来源及分布 2n=34，日本神奈川农业试验场育成，母本为天之川，父本为今村秋，1915 杂交，1927 年命名。在我国胶东半岛、黄河故道等地有栽培。

主要性状 树势较强，丰产，始果年龄早。叶片长 10.4cm，宽 6.4cm，卵圆形，初展叶绿黄色。花蕾白色，边缘淡粉色，每花序 3 ~ 6 朵花，平均 5.0 朵；雄蕊 22 ~ 28 枚，平均 25.3 枚。在湖北武汉，果实 9 月上旬成熟，单果重 484g，纵径 8.7cm，横径 9.6cm，阔圆锥形；果皮黄褐色；果心小，5 心室；果肉白色，肉质中粗，疏松，汁液多，味甜；含可溶性固形物 12.49%，可滴定酸 0.11%；品质中上等。果实较耐贮藏。

5. Niitaka

Origin and Distribution Niitaka (2n=34), originated from a cross between Amanogawa (female parent) and Imamuraaki (male parent) made by Kanagawa Agricultural Experiment Station, Japan in 1915 and released in 1927, is grown in Shandong peninsula, ancient riverbed of the Yellow River, etc.

Main Characters Tree: relatively vigorous, productive, early precocity. Leaf: 10.4cm × 6.4cm in size, ovate. Initial leaf: greenish-yellow. Flower: white bud tinged with light pink on edge, 3 to 6 flowers per cluster in average of 5.0; stamen number: 22 to 28, averaging 25.3. Fruit: matures in early September in Wuhan, Hubei Province, 484g per fruit, 8.7cm long, 9.6cm wide, broadly conical, russet skin, small core, 5 locules, flesh white, mid-coarse, tender, juicy, sweet; TSS 12.49%, TA 0.11%; quality above medium; storage life relatively long.

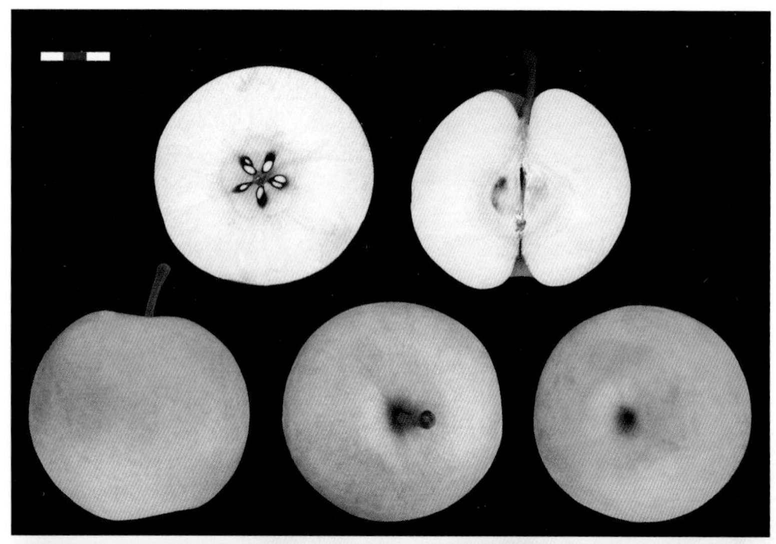

6. 二十世纪

来源及分布 2n=34，原产于日本千叶县，偶然实生。在我国山西、四川、浙江等地有少量栽培。

主要性状 树势中庸，丰产，始果年龄早或中等。叶片长 10.3cm，宽 7.4cm，卵圆形，初展叶棕红色，有茸毛。花蕾白色，每花序 3 ~ 8 朵花，平均 6.1 朵；雄蕊 27 ~ 38 枚，平均 31.5 枚。在重庆地区，果实 7 月中、下旬成熟，单果重 300g，近圆形；果皮黄绿色或黄色；果心中大，5 心室；果肉淡黄白色，肉质较细，汁液多，味甜；含可溶性固形物 11.1% ~ 14.6%；品质中上等。

6. Nijisseiki

Origin and Distribution Nijisseiki (2n=34), a chance seedling, originated in Chiba Prefecture, Japan, is grown in Shanxi, Sichuan, and Zhejiang, etc.

Main Characters Tree: moderately vigorous, productive, early or medium precocity. Leaf: 10.3cm × 7.4cm in size, ovate. Initial leaf: brownish-red. Flower: white bud, 3 to 8 flowers per cluster in average of 6.1; stamen number: 27 to 38, averaging 31.5. Fruit: matures in mid or late July in Chongqing, 300g per fruit, sub-round, yellowish-green or yellow skin, medium core, 5 locules, flesh pale yellowish-white, relatively fine, juicy, sweet; TSS 11.1%-14.6%; quality above medium.

7. 晚三吉

来源及分布 2n=34，产于日本新泻县，为早生三吉偶然实生。在我国山东、辽宁等地有栽培。

主要性状 树势中庸，较丰产，始果年龄早，自花结实率高。叶片长10.9cm，宽7.1cm，卵圆形，初展叶棕红色，微显绿色。花蕾白色，小花蕾边缘浅粉红色，每花序4～8朵花，平均6.0朵；雄蕊20枚；花冠直径4.2cm。在辽宁兴城，果实9月中、下旬成熟，单果重393g，纵径8.8cm，横径8.6cm，近圆形；果皮绿褐色或锈褐色；果心小，5心室；果肉白色，肉质细，松脆，汁液多，味甜酸；含可溶性固形物12.9%；品质上等或中上等。

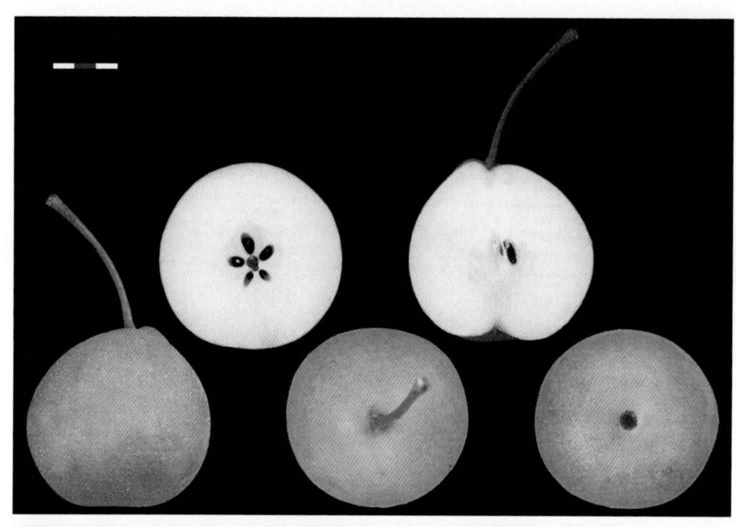

7. Okusankichi

Origin and Distribution Okusankichi (2n=34), a chance seedling of Wasesankichi, originated in Niigata Prefecture, Japan, is grown in Shandong, Liaoning, etc.

Main Characters Tree: moderately vigorous, relatively productive, early precocity, self-fertile. Leaf: 10.9cm × 7.1cm in size, ovate. Initial leaf: brownish-red with less green. Flower: white bud, small one tinged with light pink on edge, 4 to 8 flowers per cluster in average of 6.0; stamen number: 20; corolla diameter: 4.2cm. Fruit: matures in mid or late September in Xingcheng, Liaoning Province, 393g per fruit, 8.8cm long, 8.6cm wide, sub- round, greenish-brown or russet skin, small core, 5 locules, flesh white, fine, crisp tender, juicy, sweet-sour; TSS 12.9%; good or above medium in quality.

8. 新世纪

来源及分布 2n=34，日本冈山县农业试验场1945年育成，母本为二十世纪，父本为长十郎。在我国浙江、山东、福建等省有栽培。

主要性状 树势较强，树冠紧凑，丰产，始果年龄极早。叶片长9.6cm，宽6.6cm，阔卵圆形，初展叶绿色。花蕾白色，每花序4～7朵花，平均5.3朵；雄蕊27～36枚，平均30.3枚；花冠直径3.0cm。在日本冈山县，果实8月中旬成熟，较二十世纪早15d；单果重300g，近圆形，稍扁；果面平滑；果皮黄绿色或黄白色；果心中大，5心室；果肉白色，肉质较细，松脆，汁液多，味淡甜；含可溶性固形物11.0%；品质中上等。

8. Shinseiki

Origin and Distribution Shinseiki (2n=34), originated from a cross between Nijisseiki (female parent) and Choujuurou (male parent) by the Okayama Prefecture Agricultural Experiment Station, Japan in 1945, is grown in Zhejiang, Shandong, and Fujian, etc.

Main Characters Tree: relatively vigorous, compact, productive, very early precocity. Leaf: 9.6cm × 6.6cm in size, broadly ovate. Initial leaf: green. Flower: white bud, 4 to 7 flowers per cluster in average of 5.3; stamen number: 27 to 36, averaging 30.3; corolla diameter: 3.0cm. Fruit: matures in mid August in Okayama Prefecture, Japan, 15 days earlier than Nijisseiki, 300g per fruit, sub-round, slight oblate, surface smooth, yellowish-green or yellowish-white skin, medium core, 5 locules, flesh white, relatively fine, crisp tender, juicy, light sweet; TSS 11.0%; quality above medium.

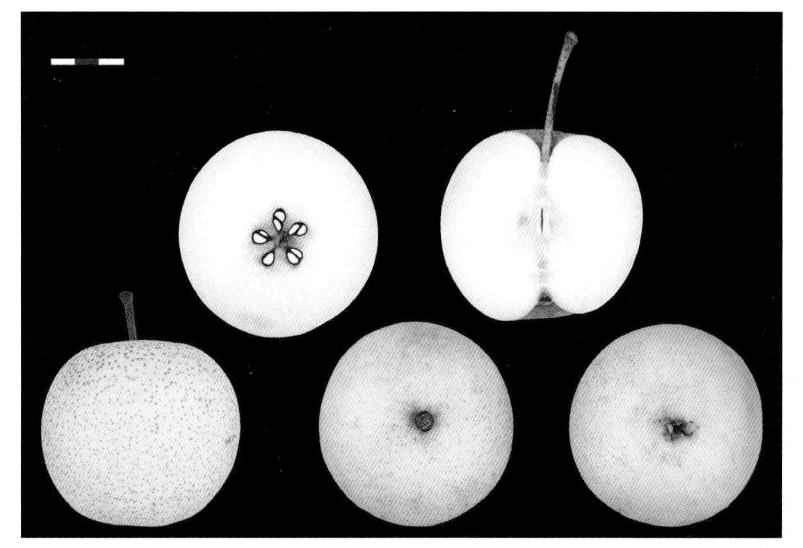

9. 新水

来源及分布 2n=34，日本农林省园艺试验场育成，1965 年命名，母本为菊水，父本为君塚早生。在江苏、上海等地有少量栽培。

主要性状 树势较强，产量中等，始果年龄早或中等。叶片长 11.9cm，宽 7.0cm，卵圆形，初展叶棕红色，微显绿色，有茸毛。花蕾白色，每花序 7 ～ 8 朵花，平均 7.3 朵；雄蕊 30 ～ 38 枚，平均 33.5 枚；花冠直径 4.4cm。在日本神奈川县，果实 8 月上旬成熟，比幸水早 7 ～ 10d；单果重 250g，扁圆形；果皮浅褐色；果心中大，5 ～ 7 心室；果肉白色，肉质细，松软，汁液多，味浓甜；含可溶性固形物 12.2%；品质上等。

9. Shinsui

Origin and Distribution Shinsui (2n=34), originated from a cross between Kikusui (female parent) and Kimizukawase (male parent) by Horticultural Experiment Station, the Ministry of Agriculture, Forestry of Japan and named in 1965, is grown limitedly in Jiangsu, Shanghai, etc.

Main Characters Tree: relatively vigorous, moderately productive, early or medium precocity. Leaf: 11.9cm × 7.0cm in size, ovate. Initial leaf: brownish-red with less green, pubescent. Flower: white bud, 7 to 8 flowers per cluster in average of 7.3; stamen number: 30 to 38, averaging 33.5; corolla diameter: 4.4cm. Fruit: matures in early August in Kanagawa Prefecture, Japan, 7 to 10 days earlier than Kousui, 250g per fruit, oblate, light brown skin, medium core, 5 to 7 locules, flesh white, fine, soft tender, juicy, sweet; TSS 12.2%; quality good.

四、韩国砂梨品种

Korean Sand Pear Varieties

1. 甘川梨

来源及分布 韩国农村振兴厅园艺研究所1990年育成，母本为晚三吉，父本为甜梨。

主要性状 树势强，丰产，始果年龄早。叶片长12.0cm，宽8.7cm，卵圆形，初展叶红色，带有绿色。花蕾白色或边缘浅粉红色，每花序5～8朵花，平均6.0朵；雄蕊20枚；花冠直径3.8cm。在韩国水原，果实10月上、中旬成熟；在北京地区，果实9月下旬成熟。单果重590g，果实阔圆锥形；果皮黄褐色或橙褐色；果心小，4或5心室；果肉白色，肉质细，松脆，味甘甜；含可溶性固形物13.3%；品质上等。果实较耐贮藏。

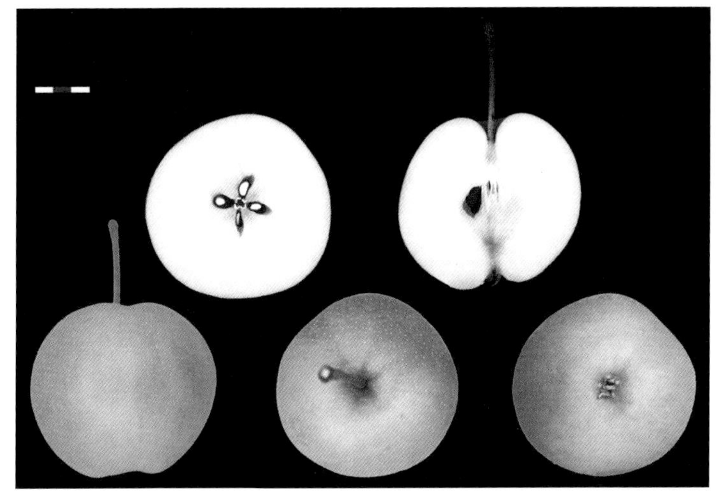

1. Gamcheonbae

Origin and Distribution Gamcheonbae, was originated from a cross between Okusankichi (female parent) and Danbae (male parent) by the National Horticultural Research Institute, R.D.A. REP. of Korea and released in 1990.

Main Characters Tree: vigorous, productive, early precocity. Leaf: 12.0cm × 8.7cm in size, ovate. Initial leaf: red with green. Flower: white bud or tinged with light pink on edge, 5 to 8 flowers per cluster in average of 6.0; stamen number: 20; corolla diameter: 3.8cm. Fruit: matures in early or mid October in Suwon, Korea, and in late September in Beijing, 590g per fruit, broadly conical, yellowish-brown or orange brown skin, small core, 4 or 5 locules, flesh white, fine, crisp tender, very sweet; TSS 13.3%; quality good; storage life relatively long.

2. 晚秀

来源及分布　韩国农村振兴厅园艺研究所 1995 年育成，母本为甜梨，父本为晚三吉。在山东、河北等地有少量栽培。

主要性状　树势强，丰产，始果年龄早。叶片长 10.3cm，宽 7.0cm，卵圆形，初展叶暗红色。花蕾白色，边缘淡粉红色，每花序 4 ～ 6 朵花，平均 4.9 朵；雄蕊 20 ～ 21 枚，平均 20.2 枚；花冠直径 3.8cm。在韩国水原，果实 10 月下旬成熟；在山东莱西，果实 10 月下旬成熟。单果重 660g，扁圆形，果皮绿褐色或黄褐色；果心小，4 或 5 心室；果肉白色，肉质较细，松脆或疏松，汁液多，味甜；含可溶性固形物 12.4%；品质上等。果实耐贮藏。

2. Mansoo

Origin and Distribution Mansoo, originated from a cross between Danbae (female parent) and Okusankichi (male parent) by the National Horticultural Research Institute, R.D.A. REP. of Korea and released in 1995, is grown limitedly in Shandong, Hebei, etc.

Main Characters Tree: vigorous, productive, early precocity. Leaf: 10.3cm × 7.0cm in size, ovate. Initial leaf: dark red. Flower: white bud tinged with light pink on edge, 4 to 6 flowers per cluster in average of 4.9; stamen number: 20 to 21, averaging 20.2; corolla diameter: 3.8cm. Fruit: matures in late October in Suwon, Korea, also late October in Laixi City, Shandong Province, 660g per fruit, oblate, greenish-brown or yellowish-brown skin, small core, 4 or 5 locules, flesh white, relatively fine, crisp tender or tender, juicy, sweet; TSS 12.4%; quality good; storage life long.

3. 黄金梨

来源及分布 韩国农村振兴厅园艺研究所 1984 年育成，母本为新高，父本为二十世纪。在我国华北平原、长江流域、黄河故道等地区有栽培。

主要性状 树势较强，丰产，始果年龄早。叶片长 11.4cm，宽 7.9cm，卵圆形，初展叶绿色。花蕾白色，每花序 5 ～ 9 朵花，平均 6.8 朵；雄蕊 20 ～ 30 枚，平均 24.7 枚；花冠直径 4.1cm。在韩国水原，果实 9 月中、下旬成熟；在北京地区，果实 9 月上旬成熟。单果重 430g，扁圆形或圆形；果皮淡黄绿色或黄色；果心小或中大，5 心室；果肉白色，肉质细，脆或松脆，汁液多，味甜；含可溶性固形物 14.9%，可滴定酸 0.13%；品质上等。

3. Whangkeumbae

Origin and Distribution Whangkeumbae, originated from a cross between Niitaka (female parent) and Nijisseiki (male parent) by the National Horticultural Research Institute, R.D.A. REP. of Korea and released in 1984, is grown along Yangtze River basin, ancient riverbed of the Yellow River and North China plain, etc.

Main Characters Tree: relatively vigorous, productive, early precocity. Leaf: 11.4cm × 7.9cm in size, ovate. Initial leaf: green. Flower: white bud, 5 to 9 flowers per cluster in average of 6.8; stamen number: 20 to 30, averaging 24.7; corolla diameter: 4.1cm. Fruit: matures in mid to late September in Suwon, Korea, and in early September in Beijing, 430g per fruit, oblate or round, light yellowish-green or yellow skin, small or medium core, 5 locules, flesh white, fine, crisp or crisp tender, juicy, sweet; TSS 14.9%, TA 0.13%; quality good.

4. 华山

来源及分布　韩国农村振兴厅园艺研究所1992年育成，母本为丰水，父本为晚三吉。在山东、河北、北京等地有少量栽培。

主要性状　树势较强，花粉量大，丰产，始果年龄早。叶片长11.7cm，宽6.9cm，卵圆形，初展叶红色，带有绿色，有茸毛。花蕾白色或边缘着浅粉红色，每花序4～7朵花，平均5.4朵；雄蕊18～28枚，平均21.4枚；花冠直径3.2cm。在韩国水原，果实9月下旬或10月上旬成熟；在山东莱西，果实9月下旬成熟。单果重580g，果实圆形或扁圆形；黄褐色，果点较大；果心中大，5心室；果肉白色，肉质细，疏松，汁液多，味甜；含可溶性固形物13.2%；品质上等。

4. Whasan

Origin and Distribution Whasan, originated from a cross between Housui (female parent) and Okusankichi (male parent) by the National Horticultural Research Institute, R.D.A. REP. of Korea and released in 1992, is grown limitedly in Shandong, Hebei, and Beijing, etc.

Main Characters Tree: relatively vigorous, productive, early precocity. Leaf: 11.7cm × 6.9cm in size, ovate. Initial leaf: red with green, pubescent. Flower: white bud or tinged with light pink on edge, 4 to 7 flowers per cluster in average of 5.4; stamen number: 18 to 28, averaging 21.4; corolla diameter: 3.2cm. Fruit: matures in late September or early October in Suwon, Korea, and late September in Laixi City, Shandong Province, 580g per fruit, round or oblate, yellowish-brown skin with dots relatively large and obvious, medium core, 5 locules, flesh white, fine, juicy, tender, sweet; TSS 13.2%; quality good.

5. 圆黄

来源及分布 韩国农村振兴厅园艺研究所1994年育成，母本为早生赤，父本为晚三吉。在我国北京、河北、山东、四川等地有栽培。

主要性状 树势强，丰产，始果年龄早。叶片长11.7cm，宽7.0cm，卵圆形，初展叶绿色。花蕾白色，每花序3～7朵花，平均4.7朵；雄蕊19～23枚，平均21.3枚。在韩国水原，果实9月上、中旬成熟；在山东莱西，果实9月上旬成熟。单果重570g，圆形或扁圆形；果皮黄褐色；果心中大，5心室；果肉淡黄白色，肉质极细，松脆或疏松，汁液极多，味甘甜；含可溶性固形物13.4%，可滴定酸0.14%；品质上等或极上。

5. Wonhwang

Origin and Distribution Wonhwang, originated from a cross between Waseaka (female parent) and Okusankichi (male parent) by the National Horticultural Research Institute, R.D.A. REP. of Korea and released in 1994, is grown in Beijing, Hebei, Shandong, and Sichuan, etc.

Main Characters Tree: vigorous, productive, early precocity. Leaf: 11.7cm × 7.0cm in size, ovate. Initial leaf: green. Flower: white bud, 3 to 7 flowers per cluster in average of 4.7; stamen number: 19 to 23, averaging 21.3. Fruit: matures in early or mid September in Suwon, Korea, and early September in Laixi City, Shandong Province, 570g per fruit, round or oblate, yellowish-brown skin, medium core, 5 locules, flesh pale yellowish-white, extremely fine and juicy, crisp tender or tender, very sweet; TSS 13.4%, TA 0.14%; good or extremely good in quality.

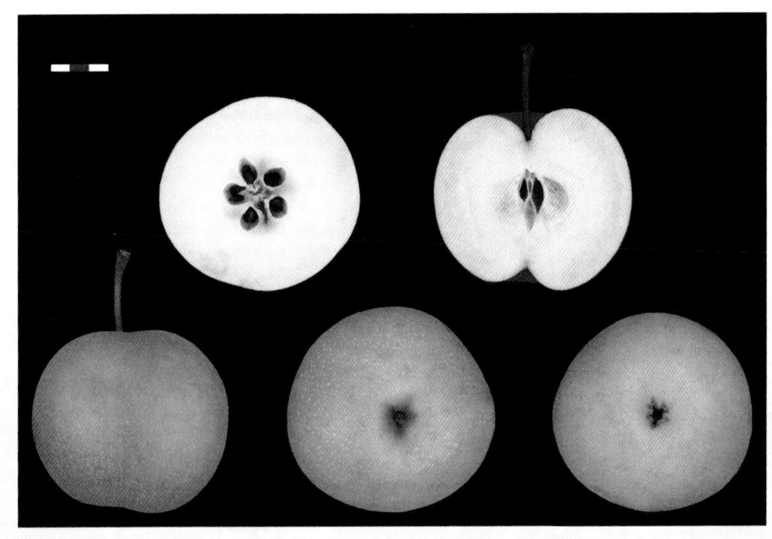

五、西洋梨品种

Common Pear Varieties

1. 阿巴特

来源及分布 法国品种，1866年发现，来源不详。在我国胶东半岛、北京等地有栽培。

主要性状 树势中庸，丰产，始果年龄早。叶片长7.4cm，宽4.1cm，椭圆形，初展叶绿色。花蕾白色，每花序5～7朵花，平均6.1朵；雄蕊14～23枚，平均19.2枚；花冠直径4.0cm。在山东烟台，果实9月上旬成熟。单果重310g，长颈葫芦形；果皮绿色或黄色，部分果实阳面有红晕；果心小，5或4心室；果肉白色，肉质细，采收后即可食用，经10～12d后熟，肉质变软或软溶，汁液多，味甜，有香气；含可溶性固形物13.0%；品质上等。

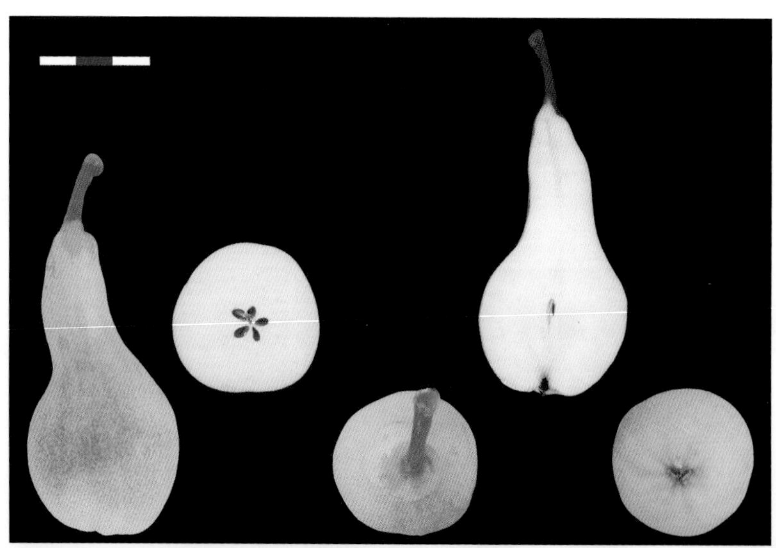

1. Abbe Fetel

Origin and Distribution Abbe Fetel (Abate Fetel), originated in France in 1866, which pedigree is unknown, is cultivated in Shandong Peninsula and Beijing, etc.

Main Characters Tree: moderately vigorous, productive, early precocity. Leaf: 7.4cm × 4.1cm in size, elliptical. Initial leaf: green. Flower: white bud, 5 to 7 flowers per cluster in average of 6.1; stamen number: 14 to 23, averaging 19.2; corolla diameter: 4.0cm. Fruit: matures in early September in Yantai, Shandong Province, 310g per fruit, elongated pyriform, green or yellow skin, some covered with light red on the side exposed to the sun, small core, 5 or 4 locules, flesh white, fine, edible after harvest, becoming soft or melting after 10 to 12 days of storage, juicy, sweet, aromatic; TSS 13.0%; quality good.

2. 巴梨

来源及分布 2n=34，原产英国，1770年在英国伯克郡发现，偶然实生。在我国胶东半岛、辽东半岛、黄河故道等地有栽培。

主要性状 树势中庸，丰产，始果年龄中等。叶片长6.6cm，宽4.2cm，椭圆形，初展叶黄绿色。花蕾白色，小花蕾边缘淡粉红色，每花序5～8朵花，平均6.6朵；雄蕊20～24枚，平均21.2枚；花冠直径3.4cm。在辽宁兴城，果实8月下旬至9月上旬成熟，单果重217g，纵径9.8cm，横径7.0cm，粗颈葫芦形，不规则，表面凹凸不平；果皮绿黄色或黄色，少数果实阳面有红晕；果心较小，5心室；果肉白色，肉质细，经7～10d后熟，肉质变软溶，汁液特多，味酸甜，有浓香；含可溶性固形物13.85%，可滴定酸0.28%；品质极上。果实不耐贮藏。

2. Bartlett

Origin and Distribution Bartlett (2n=34), a chance seedling, discovered in Berkshire, England in 1770, is grown mainly in Shandong Peninsula, Liaodong Peninsula, and ancient riverbed of the Yellow River, etc.

Main Characters Tree: moderately vigorous, productive, medium precocity. Leaf: 6.6cm × 4.2cm in size, elliptical. Initial leaf: yellowish-green. Flower: white bud, small one tinged with light pink on edge, 5 to 8 flowers per cluster in average of 6.6; stamen number: 20 to 24, averaging 21.2; corolla diameter: 3.4cm. Fruit: matures in late August or early September in Xingcheng, Liaoning Province, 217g per fruit, 9.8cm long, 7.0cm wide, obtuse-pyriform, irregular, surface slightly uneven, greenish-yellow or yellow skin, few covered with red on the side exposed to the sun, relatively small core, 5 locules, flesh white, fine, melting after 7 to 10 days of storage, extremely juicy, sour-sweet, strongly aromatic; TSS 13.85%, TA 0.28%; excellent in quality; storage life short.

3. 红巴梨

来源及分布 2n=34，原产美国，1945 年发现，为巴梨芽变。在胶东半岛、辽东半岛、甘肃等地有少量分布。

主要性状 树势中庸，较丰产，始果年龄早。叶片长 7.5cm，宽 4.2cm，椭圆形，初展叶黄绿色，着浅红色。花蕾白色，边缘浅粉红色，每花序 6 ~ 7 朵花，雄蕊 20 ~ 23 枚。在辽宁兴城，果实 9 月上旬成熟，单果重 229g，纵径 10.0cm，横径 7.3cm，粗颈葫芦形；果皮绿黄色，阳面有红晕；果心较小，5 心室；果肉白色，肉质细，紧密，后熟变软溶，汁液多，味甜，有香气；含可溶性固形物 13.68%；品质上等。

3. Bartlett - Max Red

Origin and Distribution Bartlett - Max Red (2n=34), a bud mutation of Bartlett, originated in the United States in 1945, is grown limitedly in Shandong Peninsula, Liaodong Peninsula, and Gansu, etc.

Main Characters Tree: moderately vigorous, relatively productive, early precocity. Leaf: 7.5cm × 4.2cm in size, elliptical. Initial leaf: yellowish-green tinged with light red. Flower: white bud tinged with light pink on edge, 6 to 7 flowers per cluster; stamen number: 20 to 23. Fruit: matures in early September in Xingcheng, Liaoning Province, 229g per fruit, 10.0cm long, 7.3cm wide, obtuse-pyriform, greenish-yellow skin covered with red on the side exposed to the sun, relatively small core, 5 locules, flesh white, fine, dense, becoming melting after storage, juicy, sweet, aromatic; TSS 13.68%; quality good.

4. 伏茄

来源及分布 2n=34，原产法国，1825年发现，偶然实生。在我国胶东半岛及黄河故道等地有少量栽培。

主要性状 树势强，产量中等。叶片长6.9cm，宽4.4cm，椭圆形，初展叶绿色。花蕾白色，每花序5～7朵花，平均5.6朵；雄蕊20～25枚，平均21.8枚；花冠直径3.3cm。在辽宁兴城，果实7月下旬成熟，单果重82g，纵径5.2cm，横径3.0cm，葫芦形或细颈葫芦形，果柄斜生或呈弓形；果皮淡绿色，后熟转淡黄色，阳面有红晕；果心小，5心室；果肉白色，肉质细，紧密，经3～5d后熟变软溶，汁液中多，味酸甜，微有香气；含可溶性固形物15.17%，可滴定酸0.32%；品质中上或上等。果实不耐贮藏。

4. Beurré Giffard

Origin and Distribution Beurré Giffard (2n=34), a chance seedling, originated in France in 1825 , is grown limitedly in Shandong Peninsula and ancient riverbed of the Yellow River, etc.

Main Characters Tree: vigorous, medium production. Leaf: 6.9cm × 4.4cm in size, elliptical. Initial leaf: green. Flower: white bud, 5 to 7 flowers per cluster in average of 5.6; stamen number: 20 to 25, averaging 21.8; corolla diameter: 3.3cm. Fruit: matures in late July in Xingcheng, Liaoning Province, 82g per fruit, 5.2cm long, 3.0cm wide, pyriform or narrow-necked pyriform, stalk inclined or arched, light green skin covered with red on the side exposed to the sun, turning to pale yellow after storage, small core, 5 locules, flesh white, dense, turning to melting after 3 to 5 days of storage, fine, mid-juicy, sour-sweet, slightly aromatic; TSS 15.17%, TA 0.32%; good or above medium in quality; storage life short.

5. 哈代

来源及分布 原产法国，1820 年发现，来源不详。常嫁接到榅桲上作为西洋梨的中间砧利用，表现半矮化。

主要性状 树势强，丰产。叶片长 8.7cm，宽 5.7cm，椭圆形，初展叶绿色。花蕾白色，每花序 6 ~ 9 朵花，平均 7.4 朵；雄蕊 19 ~ 26 枚，平均 21.5 枚；花冠直径 3.6cm。在辽宁兴城，果实 8 月底成熟，单果重 130g，纵径 7.4cm，横径 6.2cm，葫芦形或短葫芦形；果皮黄绿色，有果锈；果心小，5 心室；果肉白色，肉质细，紧密，后熟变软，汁液多，味酸甜，有香气；含可溶性固形物 13.85%；品质中上等。

5. Beurré Hardy

Origin and Distribution Beurré Hardy, pedigree unknown, a semi-dwarf on quince which was often used as an intermediate trunk stock for Common Pear trees, was originated in France in 1820.

Main Characters Tree: vigorous, productive. Leaf: 8.7cm × 5.7cm in size, elliptical. Initial leaf: green. Flower: white bud, 6 to 9 flowers per cluster in average of 7.4; stamen number: 19 to 26, averaging 21.5; corolla diameter: 3.6cm. Fruit: matures in late August in Xingcheng, Liaoning Province, 130g per fruit, 7.4cm long, 6.2cm wide, pyriform or obtuse-pyriform, yellowish-green skin, often with some russet, small core, 5 locules, flesh white, fine, dense, turning to soft after storage, juicy, sour-sweet, aromatic; TSS 13.85%; quality above medium.

6. 车头梨

来源及分布 又名朝鲜洋梨，可能由朝鲜引入，从植株外形看，似西洋梨与砂梨杂交品种。在我国吉林延边等地有栽培。

主要性状 树势强，较丰产，始果年龄中或晚。叶片长 10.3cm，宽 5.8cm，卵圆形，初展叶绿色。花蕾白色，小花蕾边缘浅粉红色，每花序 6 ~ 9 朵花，平均 7.0 朵；雄蕊 23 ~ 27 枚，平均 24.3 枚；花冠直径 4.1cm。在辽宁兴城，果实 8 月中、下旬成熟，单果重 170g，纵径 6.0cm，横径 6.8cm，扁圆形，果面平滑；果皮绿黄色，少数果实阳面有红晕；果心中大，5 心室；果肉白色，肉质细，脆，经 5 ~ 7d 后熟，肉质变松软，汁液多，味酸甜；含可溶性固形物 12.63%，可滴定酸 0.28%；品质中上等。果实不耐贮藏。

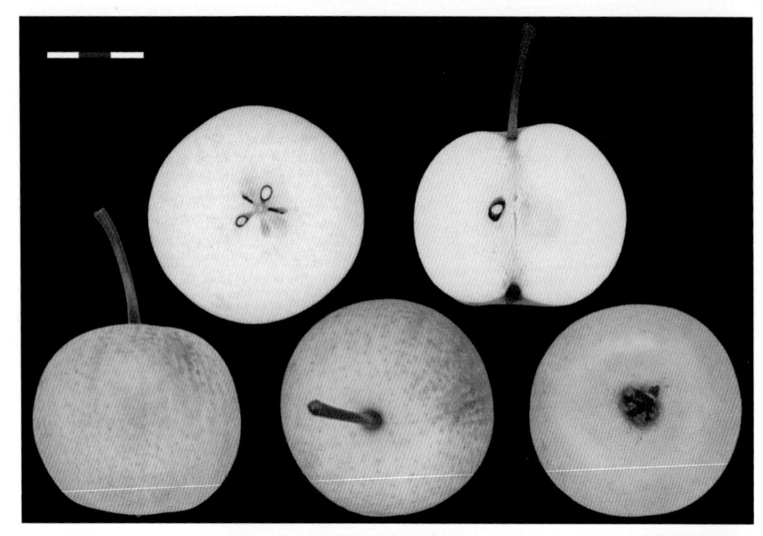

6. Chetouli

Origin and Distribution Chetouli, known as Korean Common Pear, which might be an inter-specific hybridization cultivar between Common Pear and Sand Pear varieties seen from its characters, might be introduced from Korea, is grown in Yanbian, Jilin Province, etc.

Main Characters Tree: vigorous, relatively productive, medium or late precocity. Leaf: 10.3cm × 5.8cm in size, ovate. Initial leaf: green. Flower: white bud, small bud tinged with light pink on edge, 6 to 9 flowers per cluster in average of 7.0; stamen number: 23 to 27, averaging 24.3; corolla diameter: 4.1cm. Fruit: matures in mid or late August in Xingcheng, Liaoning Province, 170g per fruit, 6.0cm long, 6.8cm wide, oblate, surface smooth, greenish-yellow skin, few covered with red on the side exposed to the sun, medium core, 5 locules, flesh white, fine, crisp, becoming tender soft after 5 to 7 days of storage, juicy, sour-sweet; TSS 12.63%, TA 0.28%; quality above medium; storage life short.

7. 茄梨

来源及分布 原产美国，母本为日面红，父本为巴梨，1860年育成。在我国胶东半岛、黄河故道等地有栽培。

主要性状 树势强，丰产，始果年龄中或晚，抗寒性较强。叶片长8.8cm，宽5.4cm，椭圆形，初展叶绿色。花蕾白色，边缘粉红色，每花序5～8朵花，平均6.9朵；雄蕊23～26枚，平均24.1枚；花冠直径3.8cm。在辽宁兴城，果实8月下旬成熟，较巴梨早7～10d。单果重175g，纵径8.2cm，横径6.8cm，短葫芦形或倒卵圆形；果皮绿黄色，阳面有淡红晕；果心中大，5心室或4心室；果肉白色，肉质较细，紧密，经7～10d后熟，肉质变软溶，汁液多，味甜，有香气；含可溶性固形物12.75%，可滴定酸0.23%；品质上等。

7. Clapp Favorite

Origin and Distribution Clapp Favorite (2n=34), originated from a cross between Flemish Beauty (female parent) and Bartlett (male parent) in the United States in 1860, is grown in Shandong Peninsula, ancient riverbed of the Yellow River, etc.

Main Characters Tree: vigorous, productive, medium or late precocity, relatively hardy. Leaf: 8.8cm × 5.4cm in size, elliptical. Initial leaf: green. Flower: white bud tinged with pink on edge, 5 to 8 flowers per cluster in average of 6.9; stamen number: 23 to 26, averaging 24.1; corolla diameter: 3.8cm. Fruit: matures in late August in Xingcheng, Liaoning Province, 7 to 10 days earlier than Bartlett, 175g per fruit, 8.2cm long, 6.8cm wide, obtuse-pyriform or obovate, greenish-yellow skin covered with light red on the side exposed to the sun, medium core, 5 or 4 locules, flesh white, relatively fine, dense, turning to melting after 7 to 10 days of storage, juicy, sweet, aromatic; TSS 12.75%, TA 0.23%; quality good.

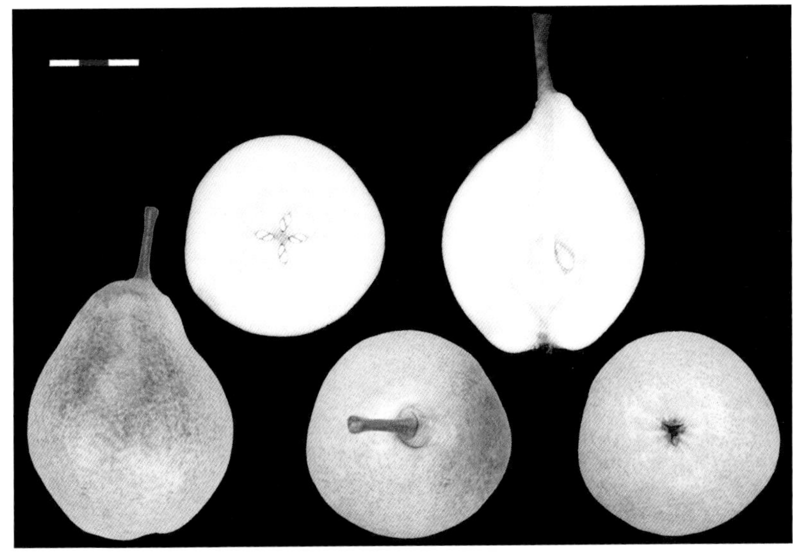

8. 康佛伦斯

来源及分布 2n=34，英国品种，Leon Leclerc de Laval 实生，1894 年发现。在我国山东烟台等地有少量栽培。

主要性状 树势中庸，丰产，始果年龄中或晚，自花授粉结实率较高。叶片长 7.8cm，宽 4.9cm，椭圆形，初展叶绿色。花蕾白色，每花序 6 ～ 9 朵花，平均 7.1 朵；雄蕊 20 ～ 25 枚，平均 21.0 枚；花冠直径 3.5cm。在山东烟台，果实 9 月上旬成熟，单果重 300g，细颈葫芦形；果皮绿色，不规则着生果锈；果心较小，5 心室；果肉白色，肉质细，紧密而脆，经后熟变软，肉质软溶，汁液多，味甜，有香气；含可溶性固形物 12.45%；品质上等。

8. Conference

Origin and Distribution Conference (2n=34), a seedling of Leon Leclerc de Laval, originated in the United Kingdom in 1894, is grown limitedly in Yantai, Shandong Province, etc.

Main Characters Tree: moderately vigorous, productive, medium or late precocity, self-fertile. Leaf: 7.8cm × 4.9cm in size, elliptical. Initial leaf: green. Flower: white bud, 6 to 9 flowers per cluster in average of 7.1; stamen number: 20 to 25, averaging 21.0; corolla diameter: 3.5cm. Fruit: matures in early September in Yantai, Shandong Province, 300g per fruit, narrow-necked pyriform, green skin, unevenly russet, relatively small core, 5 locules, flesh white, fine, dense and crisp, turning to melting after storage, juicy, sweet, aromatic; TSS 12.45%; quality good.

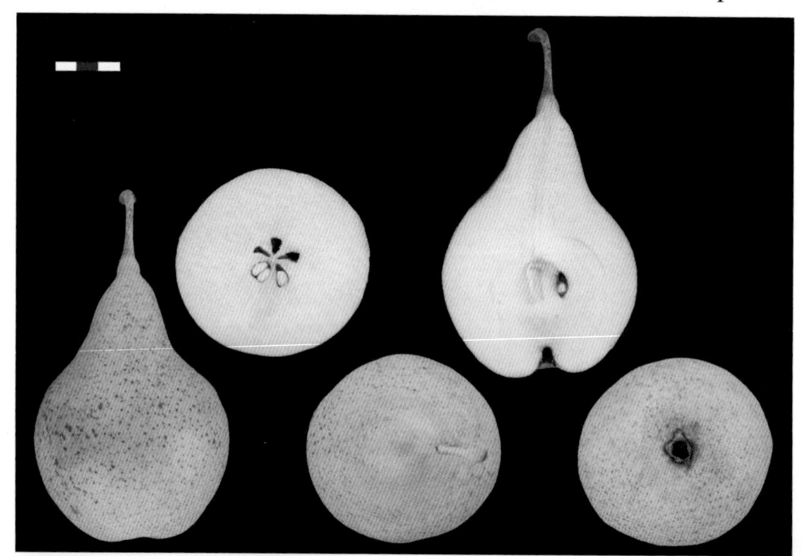

9. 三季梨

来源及分布 2n=34，法国品种，1870 年发现，实生。在我国胶东半岛、辽东半岛、辽西等地有栽培。

主要性状 树势中庸，较丰产，始果年龄中或晚。叶片长 6.3cm，宽 4.4cm，椭圆形，初展叶绿色。花蕾白色，每花序 4 ~ 8 朵花，平均 5.7 朵；雄蕊 22 ~ 30 枚，平均 25.0 枚。在辽宁兴城，果实 8 月下旬成熟，成熟期较巴梨稍早。单果重 249g，纵径 10.5cm，横径 7.1cm，粗颈长葫芦形，果面凹凸不平；果皮绿黄色，部分果实阳面有红晕；果心较小，5 或 6 心室；果肉白色，较细，紧密，经 10d 左右后熟，肉质变软或软溶，易沙面，汁液多或中多，味酸甜，有香气；含可溶性固形物 11.48%；品质中上等。果实不耐贮藏。

9. Docteur Jules Guyot

Origin and Distribution Docteur Jules Guyot (2n=34), a seedling selection, originated in France in 1870, is grown in Shandong Peninsula, Liaodong Peninsula, western parts of Liaoning Province, etc.

Main Characters Tree: moderately vigorous, relatively productive, medium or late precocity. Leaf: 6.3cm × 4.4cm in size, elliptical. Initial leaf: green. Flower: white bud, 4 to 8 flowers per cluster in average of 5.7; stamen number: 22 to 30, averaging 25.0. Fruit: matures in late August in Xingcheng, Liaoning Province, some days earlier than Bartlett, 249g per fruit, 10.5cm long, 7.1cm wide, long obtuse-pyriform, irregular, greenish-yellow skin, some covered with red on the side exposed to the sun, relatively small core, 5 or 6 locules, flesh white, relatively fine, dense, turning to soft or melting after 10 days of storage, quickly becoming mealy, juicy or mid-juicy, sour-sweet, aromatic; TSS 11.48%; quality above medium; storage life short.

10. 日面红

来源及分布 2n=34，比利时品种，19 世纪初发现，偶然实生。在我国胶东半岛、辽东半岛、黄河故道、甘肃等地有栽培。

主要性状 树势强，较丰产，始果年龄晚。叶片长 7.4cm，宽 4.5cm，椭圆形，初展叶绿色。花蕾白色，边缘浅粉红色，每花序 5 ~ 7 朵花，平均 5.9 朵；雄蕊 20 ~ 31 枚，平均 25.2 枚；花冠直径 3.4cm。在辽宁兴城，果实 9 月上旬成熟，较巴梨稍晚。单果重 250g，纵径 8.5cm，横径 7.5cm，短葫芦形或倒卵圆形，果面平滑；果皮绿黄色，阳面有红晕；果心小，5 心室；果肉白色，肉质中粗，硬，经 5 ~ 7d 后熟，肉质软溶，汁液多，味甜，有香气；含可溶性固形物 15.67%，可滴定酸 0.25%；品质中上等。

10. Flemish Beauty

Origin and Distribution Flemish Beauty (2n=34), a chance seedling, originated in Belgium which was found at the beginning of the 19th century, is grown in Shandong Peninsula, Liaodong Peninsula, ancient riverbed of the Yellow River, and Gansu Province, etc.

Main Characters Tree: vigorous, relatively productive, late precocity. Leaf: 7.4cm × 4.5cm in size, elliptical. Initial leaf: green. Flower: white bud tinged with light pink on edge, 5 to 7 flowers per cluster in average of 5.9; stamen number: 20 to 31, averaging 25.2; corolla diameter: 3.4cm. Fruit: matures in early September in Xingcheng, Liaoning Province, some days later than Bartlett, 250g per fruit, 8.5cm long, 7.5cm wide, obtuse-pyriform or obovate, greenish-yellow skin, very smooth, often blushed, small core, 5 locules, flesh white, somewhat granular, firm, becoming buttery after 5 to 7 days of storage, juicy, sweet, aromatic; TSS 15.67%, TA 0.25%; quality above medium.

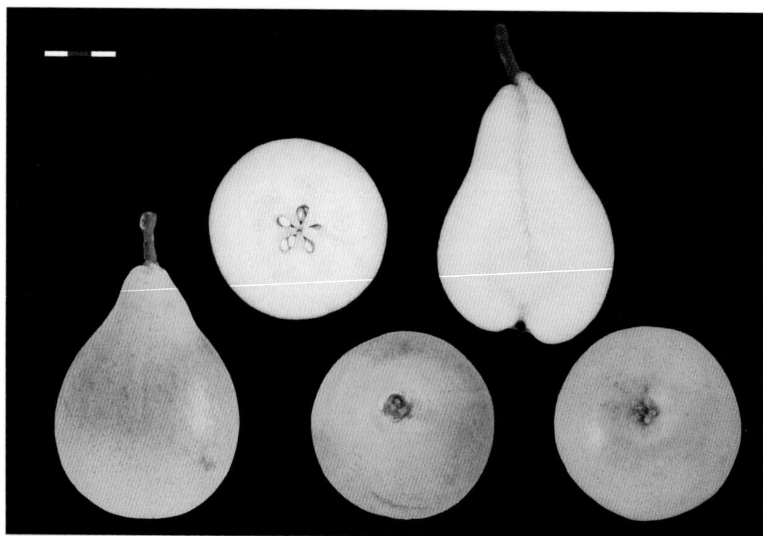

11. 派克汉姆

来源及分布 原产澳大利亚，1897 年选育，母本为 Uvedale St. Germain，父本为 Bartlett。在我国山东烟台等地有少量栽培。

主要性状 树势中庸，丰产，稳产。叶片长 7.1cm，宽 3.5cm，椭圆形，初展叶绿色。花蕾白色，每花序 5 ~ 8 朵花，平均 6.1 朵；雄蕊 20 ~ 22 枚，平均 20.4 枚；花冠直径 4.2cm。果实成熟期较巴梨晚 30d。单果重 317g，纵径 10.1cm，横径 8.3cm，粗颈葫芦形；表面稍显粗糙，凹凸不平；果皮绿黄色或黄色；果心极小，5 心室；果肉白色，肉质细，紧密，后熟变软溶，汁液多，味甜，有香气；含可溶性固形物 12.70%；品质上等。果实冷藏贮藏期与安久梨相当。

11. Packham's Triumph

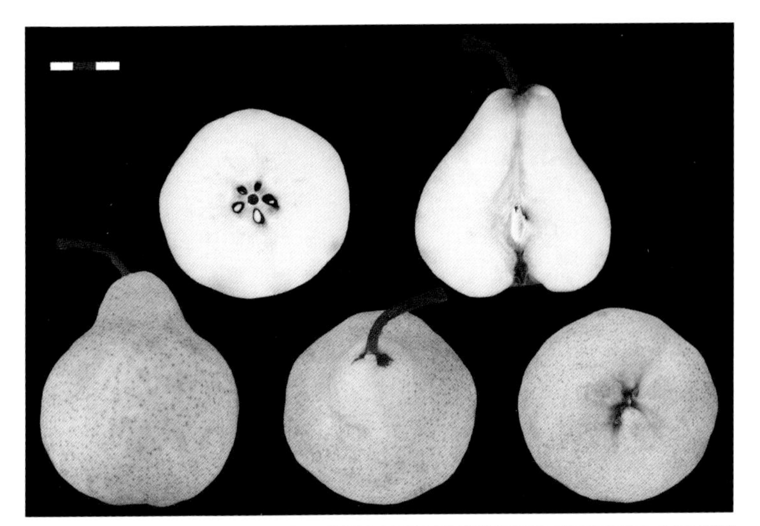

Origin and Distribution Packham's Triumph, originated in Australia in 1897, which pedigree is Uvedale St. Germain × Bartlett, is grown limitedly in Yantai, Shandong Province, etc.

Main Characters Tree: moderately vigorous, productive, stable production. Leaf: 7.1cm × 3.5cm in size, elliptical. Initial leaf: green. Flower: white bud, 5 to 8 flower per cluster in average of 6.1; stamen number: 20 to 22, averaging 20.4; corolla diameter: 4.2cm. Fruit: matures 30 days later than Bartlett, 317g per fruit, 10.1cm long, 8.3cm wide, obtuse-pyriform, surface somewhat rough, slightly uneven, greenish-yellow or yellow skin, extremely small core, 5 locules, flesh white, fine, dense, becoming melting after storage, juicy, sweet, aromatic; TSS 12.70%; quality good; storage life as long as Beurre d'Anjou in cold storage.

12. 红茄梨

来源及分布 美国品种，1950 年发现，1956 年专利注册，为茄梨的红色芽变。

主要性状 树势较强，丰产，始果年龄中或晚。叶片长 9.1cm，宽 4.1cm，椭圆形，初展叶绿色。花蕾白色或边缘浅粉红色，每花序 4 ~ 7 朵花，平均 6.2 朵；雄蕊 20 ~ 24 枚，平均 22.7 枚；花冠直径 3.7cm。在辽宁兴城，果实 8 月中、下旬成熟，单果重 131g，纵径 8.5cm，横径 6.2cm，葫芦形，果面平滑；果皮为全面深红色，美观；果心中大或小，5 或 4 心室；果肉浅黄白色，肉质紧密，经 5 ~ 7d 后熟，肉质软溶，汁液多，味甜，有香气；含可溶性固形物 12.30%，可滴定酸 0.24%；品质上等。果实不耐贮藏。

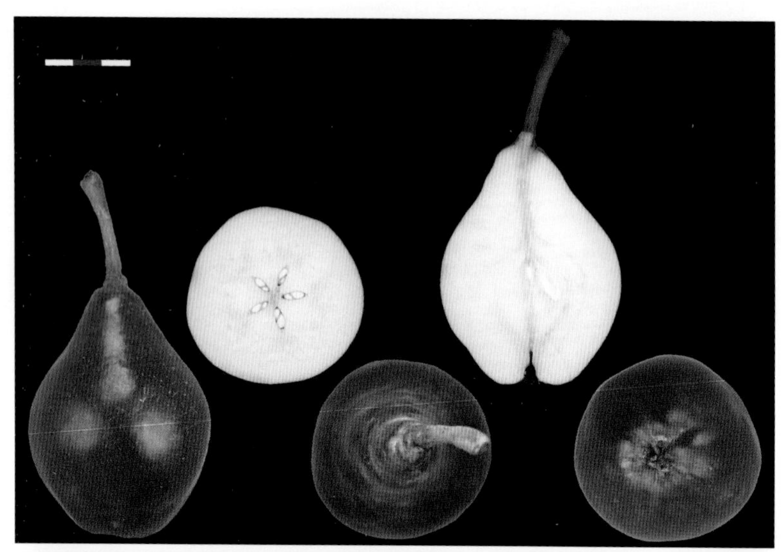

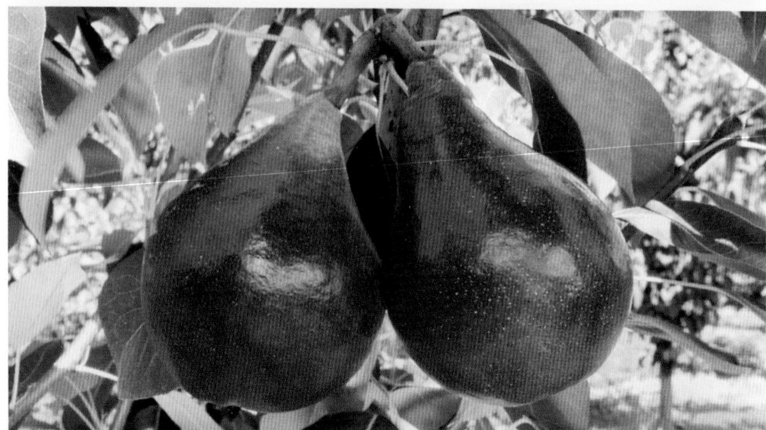

12. Red Clapp Favorite

Origin and Distribution Red Clapp Favorite (Kalle, Starkrimson), a red sport of Clapp Favorite, was originated in the United States in 1950 and patented in 1956.

Main Characters Tree: relatively vigorous, productive, medium or late precocity. Leaf: 9.1cm × 4.1cm in size, elliptical. Initial leaf: green. Flower: white bud or tinged with light pink on edge, 4 to 7 flowers per cluster in average of 6.2; stamen number: 20 to 24, averaging 22.7; corolla diameter: 3.7cm. Fruit: matures in mid or late August in Xingcheng, Liaoning Province, 131g per fruit, 8.5cm long, 6.2cm wide, pyriform, surface smooth, solid crimson red skin, very attractive, medium or small core, 5 or 4 locules, flesh pale yellowish-white, dense, becoming melting after 5 to 7 days of storage, juicy, sweet, aromatic; TSS 12.30%, TA 0.24%; quality good; storage life short.

附录 I　中国梨品种中文名称索引

Appedix I　Index to Pear Varieties in China (in Chinese)

附录II 中国梨品种英文名称索引

AppedixII Index to Pear Varieties in China (in English)

主要参考文献
Main References

曹玉芬，刘凤之，胡红菊，等. 2006. 梨种质资源描述规范和数据标准[M]. 北京：中国农业出版社.

贾敬贤. 1993. 果树种质资源目录 · 第一集[M]. 北京：中国农业出版社.

贾敬贤. 1998. 果树种质资源目录 · 第二集[M]. 北京：中国农业出版社.

蒲富慎，王宇霖. 1963. 中国果树志 · 第三卷：梨[M]. 上海: 上海科学技术出版社.

俞德浚. 1979. 中国果树分类学[M]. 北京：农业出版社.

张绍铃. 2013. 梨学[M]. 北京：中国农业出版社.

中华人民共和国农业部. 2013. 2012年农业统计资料[M]. 北京：中国农业出版社.

Bell R L, Quamme H A, Layne R E C, et al. 1996. Pears [M] //Janick J, Moore JN (eds) Fruit breeding, Volume I: Tree and tropical fruits. Wiley, New York, 441–514.

图书在版编目（CIP）数据

中国梨品种/曹玉芬主编．—北京：中国农业出版社，2014.9

ISBN 978-7-109-19532-5

Ⅰ．①中…　Ⅱ．①曹…　Ⅲ．①梨-品种-中国　Ⅳ．①S661．2

中国版本图书馆CIP数据核字（2014）第201575号

中国农业出版社出版
（北京市朝阳区麦子店街18号楼）
（邮政编码100125）
责任编辑　张　利　黄　宇

北京中科印刷有限公司印刷　新华书店北京发行所发行
2014年10月第1版　2014年10月北京第1次印刷

开本：889mm×1194mm　1/16　印张：12.75
字数：372千字
定价：200.00元